An Anthropology of Robots and AI

This book explores the making of robots in labs at the Massachusetts Institute of Technology (MIT). It examines the cultural ideas that go into the making of robots, and the role of fiction in co-constructing the technological practices of the robotic scientists. The book engages with debates in anthropological theorizing regarding the way that robots are reimagined as intelligent, autonomous and social, and woven into lived social realities. Richardson charts the move away from the "worker" robot of the 1920s to the "social" one of the 2000s, as robots are reimagined as companions, friends and therapeutic agents.

Kathleen Richardson is Senior Research Fellow in the Ethics of Robotics in the School for Computer Science and Informatics, Faculty of Technology, De Montfort University, Leicester, UK.

Routledge Studies in Anthropology

An Anthropology of Robots and AI

Annihilation Anxiety and Machines

Kathleen Richardson

Routledge
Taylor & Francis Group

NEW YORK AND LONDON

First published 2015
by Routledge
711 Third Avenue, New York, NY 10017

and by Routledge
2 Park Square, Milton Park, Abingdon, Oxon OX14 4RN

First issued in paperback 2017

Routledge is an imprint of the Taylor & Francis Group, an informa business

Library of Congress Cataloging-in-Publication Data
Library of Congress Control Number: 2015930211

ISBN 13: 978-0-8153-4646-3 (pbk)
ISBN 13: 978-1-138-83174-2 (hbk)

Typeset in Sabon
by Apex CoVantage, LLC

In loving memory of my brother Mark

Contents

Acknowledgments

First and foremost, I would like to thank Rodney Brooks, Jamie Rollins, Una-May O'Reilly (and the Le-Baron family who welcomed me into their home), Martin Martin, Brian Adams, Paul Fitzpatrick, Eduardo Torres-Jara, Jessica Howe, Charlie Kemp, Jessica Banks, Aaron Edsinger, Theresa Langston, Anne Lawthers (aka Ladybug), Annika Pfluger and Ron Wiken. I am very grateful to all the members of the lab for allowing me to come and see their fascinating work. I would like to say a special thank you to Lijin Aryananda, whose ongoing help and support I continue to receive as a very personal friend and as a professional robotic scientist. Also Max Berniker, Rebecca Bureau and Minshu Son—these were really the best of friends. Also I would like to thank all the users of the Stata Center who shared their thoughts with me about the space. I was so deeply moved by the generosity of the staff, students and faculty at MIT and made many lifelong friends.

There were many others who supported my academic studies and helped me along the way as friends, family and supporters, and these include: Stephen & Karen Jones, Michael Sun, Andy Aryananda and Popo Wiryanti. I could not have completed my studies without the crucial financial support of the following: The Jirehouse Foundation, The Economic Social Research Council and The British Academy Postdoctoral Fellowship.

I would like to thank the editors at *Etnofoor* for their kind permission in allowing me to reproduce some of the arguments in Chapter 5 from: Richardson, K 2010, 'Disabling as mimesis and alterity: making human-oid robots at the Massachusetts Institute of Technology', *Etnofoor*, vol. 22, no. 1, pp. 75–90.

I would like to thank two mentors from the anthropology department at Cambridge, Marilyn Strathern and Nikolai Ssorin-Chiakov. A special thanks to my friend Vita Peacock who is my important confidant, and new friends who offered ideas on different chapters: Floyd Codlin, Marek Sinason and Robert Oates. I would also like to thank the editors at Routledge for their support.

Finally, I would like to make a special tribute to CB who has helped me to understand the true meaning of love as a mutual and creative relationship between *I* and *Thou*.

Introduction
Annihilation Anxiety and Machines

In an extreme view, the world can be seen as only connections, nothing else.

Tim Berners-Lee, *Weaving the Web*, 1999, p. 14.

The Terminator movies (1984–2003) show examples of robots that are super-advanced intelligent machines intent on destroying humanity to assure their supremacy. *The Terminator* is significant to begin this narrative, as it is one of the most popular fictions of a robot and it carries a central theme about human destruction. Whether you look to the past of robots or the present, this enduring theme of destruction returns. I respect that there are many other kinds of robots to consider such as robot companions, robot lovers, therapeutic robots, domestic robots and others, and we will explore these different imaginings of the robot in what follows, but for now, we will focus on the theme of human annihilation by robots.

The Terminator film caused something of a stir when first released in 1984 and was seen by millions of people around the world in the first year of broadcast.[1] It features a high-profile Hollywood action actor, Arnold Schwarzenegger, who has a very unique stature; he is known for his toned muscled physique and stands at 1.88 meters or over 6 feet 2". Schwarzenegger's speech is marked by his strong Austrian-intoned English, and his speech and language are jokingly referred to as mechanical and formalistic. Some have rudely suggested he makes the "perfect" robot! While this is not the case, it is true that we take our cultural and technological models of robots from fictions. Multiple tides flow from fictions to living practices of technoscience.

A quick summary of the plot of the first *Terminator* is necessary. Set in the future of 2029 (not far away now), a super-advanced cyborg is sent to 1984 to kill Sarah O'Connor. In this dystopian future of 2029, super-intelligent machines rule the Earth, and authority over the remaining humans is maintained by killer robots. The Terminator T-800 Model 101 is sent back in time and must destroy Sarah O'Connor. Sarah is the mother of the future leader of the human rebellion. The machines figure if they can stop her child from being born, they can save themselves problems later. In this tale that twists and turns, and folds together the future, present and past, human and nonhuman into its narrative texture, the film represents an iconic Euro-American portrayal of robots as destroyers.

This popular fiction of robots has something important to tell us about the cultural theme of destruction, and more frightening themes followed in each subsequent film: *Terminator 2: Judgment Day* (1991), *Terminator 3: Rise of the Machines* (2003), and *Terminator Salvation* (2004). And the *Terminator* story is not over yet: in July 2015, *Terminator: Genesis* is scheduled for release, and speculating from the title, involves a hint at a rebirth. Only time will tell what the next installment of this robot saga has in store!

Hollywood filmmakers may receive some reward for shaping the cultural imagination of robots, but it was not them, but another more esoteric and radical avant-garde playwright to whom we must make our first tribute in recognizing the robot as a cultural entity, and a destructive one.

The first robots emerged as characters 1920s play, *R.U.R.* (Rossum's Universal Robots), written by Czech playwright Karel Čapek. The play is unique. It is the first to coin the term "robot" and features the first cultural representations of robots. *R.U.R.* is the first play of modern fiction to bring about the end of humanity as a narrative plot of complete human annihilation (Reilly 2011). This being the first work of modern fiction to do this is significant, as prior to this, only in religious tales such as the New Testament's Book of Revelation is human annihilation a central feature when the apocalypse comes.

The robot—as first given life in a text and through theatrical performance by its creator Čapek in *R.U.R.*—is a device to explore the fears of terminus in human existence brought about by mechanization, political ideologies and high modernism, and it speaks to the theme of humanity's end. Set in the tumultuous political era of the 1920s, Čapek took the idea of the factory worker one step further by inventing the robot. He created a laboring entity to work with limited subjectivity, a functionally competent laboring device. The term robot is from the Slavic term for work ("robota"), but Čapek, inspired by his artist brother Josef, drew on another meaning of the term relating to the "robota economy", an agricultural system where peasants work extra, providing for their landowners needs before their own. Robot is Czech for 'compulsory service', akin to Slav "robota", meaning 'servitude, hardship' (Merriam-Webster 1971, p. 1964).

As we reflect back on the robot in the 1920s play and the contemporary fiction of it, there is a recurring message:

BEWARE YOUR END, HUMANITY!

In which case, we must take seriously the fear of the end of the human that is circulated in robot narratives.

In this book, I will preface and interlace each chapter with tales from robotic fictions because I want to argue that robotic fictions are taken into the lived realities of robotic practices and transferred into the making of robots, returning back into those fictions. This book is a reading and appreciation of these fictions by observing the making of robots in labs at the Massachusetts Institute of Technology (MIT). MIT is a world-renowned science and technological institution, repeatedly in the top three of the world's top research institutions (QS World Rankings 2014). MIT has a presence in popular culture that has formalized its mystique. In the 1950s classic, *The Day the Earth Stood Still* (1951), an alien spacecraft lands in Washington, guarded by a robotic life form. In the panic that ensues, it is MIT scientists that the US government calls on to help "rationalize" the situation, decipher the mystery of the alien visitor and calm the American public. In American culture at least, MIT scientists stand for impersonal rationality and arguably masculine authority in the fields of science and technology.

By the time I began thinking of my fieldwork in the early 2000s, humanoid robot labs only numbered a few around the world, notably in cities in Japan, such as Tokyo and Waseda. MIT's robot lab was one of the first in the US to begin a program of making humanoid robots, funded by a generous Defense Advanced Research Projects Agency (DARPA) grant. Robots and violence are frequent bedfellows, even when the robots produced by military funding seem to have no direct application for a military purpose, such as building a robot child.

What is in a name anyway? In labs at MIT, I realized lab titles were fragile, coming and going depending on the grant or new focus of the research director. As a visiting researcher to robot labs in the US and the UK, the lab name is really an umbrella term, often for a multitude of research activities, some humanoid in focus and others not. The MIT robotics lab was set in the Artificial Intelligence Laboratory but shared the same physical space as MIT Computer Science. In 2003 these two departments merged to become The Computer Science and Artificial Intelligence Laboratory (CSAIL). All the CSAIL researchers were relocated to a new campus building, designed by architect Frank Gehry, known for his radical, geometrically distorted designs (Gilbert-Rolfe and Gehry 2002).

In keeping with issues of anthropological commitments to confidentiality, I have given the people and the robots in the lab pseudonyms. Some of these pseudonyms I have playfully taken from Čapek's play, *R.U.R.* As the robotic scientists I work with produce artifacts such as memos, scholarly dissertations, books, papers and robots, I have only referred to public material if such activities do not conflict with my initial commitment to honor the relationships with my interlocutors, many of whom are still my friends. I experienced considerable generosity from the lab group and researchers at MIT. I found their work and their lifeworlds extraordinary, and I hope some of that uniqueness is reflected in this book.

The robotics lab had taken a humanoid turn in early 2000s after its group director had been inspired to build humanlike robots after watching the film *2001: A Space Odyssey,* made in 1968 but set in the year 2001.

The lab pioneered the first sociable robot (an oxymoronic term no doubt): A robot designed for social interaction with the intended aim of developing to such an extent that its future kin would be sophisticated enough to be a companion to humans.

This book then is about the theories and technologies that go into the making of robots, as well the people who make them and how their stories and narratives feed into the machines they create.

Haraway's observation that the boundary between fiction and reality is thoroughly breached by technoscience is, of course, accurate (1991). But what I will attempt to show in these pages is that the Real is continually asserting itself in the making of robots, and there is a sphere outside cultural constructions that has its own separate properties. The Real is the boundary. Robotics in its own ways is confronted by its own realities. When constructing models of the mechanical human, theory and practice become intertwined in distinctive, sometimes unpredictable ways. In these labs, the robotic scientists continually referenced robotic fictions when producing robots, and the robots were repeatedly meeting the constraints of the Real: the physical, social and cultural environments that acted as containers. The Real and the fictional played off against each other in unusual ways, most notably in how the theme of robot destruction was addressed by these researchers. The cultural image of the threatening robot informed the making of the robots in the lab. The following information was provided on a robotics lab website at MIT:

> Q: Are you ever worried that your robot might get 'too intelligent' or 'too powerful'?
> A: No—we have programmed the robot to spare our lives in the event that it ever attempts to organize its brethren in a bloody revolution against the human race.
>
> (MIT Humanoid Robotics Group n.d.)

Here the theme of destruction is taken up and diffused in a light-hearted way, but robots and artificial intelligence (AI) threats are present and receive more than a passing dismissal as will become apparent in the pages that follow.

ANNIHILATION ANXIETY: TO REDUCE TO NOTHING

The last few decades of anthropological theorizing have been beset by a number of theoretical problems that have resisted the dualistic analytical consequence of Cartesian dualisms and the ways these constructions have played themselves out in the construction of what life is (Latour 1993, Haraway 1991). One may say that anthropology as a discipline has suffered (and overcome) a kind of separation anxiety—about how to describe, resolve and explain dichotomous relations including those between: persons and things (Gell 1998; Strathern 1988), humans and machines (Haraway 1991; Haraway 2003; Suchman 2006; Hicks 2002; Rabinow 2011), humans and animals (Haraway 2003; Haraway 1991; Ingold 2012), the body and the mind (Csordas 1999; Featherston & Burrows 1995), humans and nonhumans (Latour 1993; Latour 2005), fact and fiction (Haraway 1991; Graham 2002), and public and private spaces (Buchli and Lucas 2001; Buchli 1997).

If anthropology is said to have dealt with and overcome separation anxiety, why is the theme of human terminus brought about by machines a persistent and recurring theme in contemporary Euro-American cultural life? Latour (1993) takes this one step further and proposes that underscoring the fear of machines is a result of asymmetrical humanism (separation anxiety):

> How could the anthropos be threatened by machines? It has them, it has put itself into them, it has divided up its own members among their members, it has built its own body with them. How could it be threatened by objects? They have all been quasi-subjects circulating within the collective they traced. It is made of them as much as they are made of it.
>
> (1993, p. 138)

For Latour, the fear of the machine is an outcome of artificially separated categories, and this is reflected in the fear of objects (robots, viruses, supercomputers or meteors) that possess autonomy and can come back and haunt humanity as detached other.

Could the fear of the machines really be an outcome of 'asymmetrical humanism', as Latour proposes? I want to suggest that fear of robots and machines is the outcome of symmetrical anti-humanism, where humans and nonhumans are placed on a par, and the human is ascribed no distinctive quality over other agents—where human agents are reduced to nothing. This is presented as an anthropological emphasis on process in the absence of ontological difference. The robot has historically been a way to talk about dehumanization and the elevation of the nonhuman. The first meanings of the robot were primarily about dehumanization, and hence Čapek's robots were human, made of flesh, blood, bones and veins, but assembled on a mechanical production line with a scientific formula (2004, p. 13). It was other artists in the 1920s that took the robot character from the play and turned it into a machine. We can look to the robot in its historical sense and its contemporary manifestations in labs and in fictions to explore these points further. I frame this recurring fear, in contrast to separation anxiety as annihilation anxiety.

Annihilation anxieties are produced by an analytical position that rejects ontological separations, combined with radical anti-essentialism—when humans and nonhumans become comparable. The dystopian horror presented in *R.U.R.* and *The Terminator* films relate to a fear of terminus, but the fear of the robot uprising is an existential fear about the end of the human (though with $8 billion in forecasted funding for military robots by 2016 (ABIResearch 2011), the physical threat of destruction is not so fictional). The robot is a way to reflect on the violence of World War I and the unprecedented destruction of human life mediated by machines. The end of the human then is intimately related to violence: death is the ultimate end of the human.

What does annihilation mean? Annihilation is one of those terms encompassing multiple meanings, and I call upon all of those meanings in proposing an analytical framework to make sense of robots and AI systems. On the one hand, annihilation means the 'act of annihilating' or 'state of being annihilated' (Webster's Third New International 1971, p. 87). We are already familiar with Euro-American narratives of technological revenge—in the form of Frankenstein's monster from the nineteenth-century classic tale by Mary Shelly (1969), or *The Matrix* Trilogy (1999–2003), where humans are imagined as batteries for AI systems. Annihilation also means 'cessation of being: NOTHINGNESS' (Webster's Third New International 1971, p. 87). Annihilation is derived from the verb "annihilate" ('ending'), and the Latin "annihilates", past participle of "annihilare": 1. 'to cause to be of no effect', 2. 'to look upon as nothing', 3. 'to reduce to nothing' (Webster's Third New International 1971, p. 87). These meanings open up another way of reflecting on endings and nothingness. Central to this discourse on robots is to highlight the reduction of the human to nothing, as a nondistinct agent in anthropological theorizing. 'To reduce to nothing' is also about the erasing of differences between humans and nonhumans. As anthropological theorizing takes an 'ontological turn' shaped by 'actor-networks', 'assemblages', 'meshwork', and 'companion species', so too are the human and nonhuman interconnected, even enmeshed with each other (Latour 2005, Rabinow 2011, Ingold 2012, Haraway 2003). Aside from the meanings the term annihilation possesses in popular language, it has meanings in physics, too, which are worth considering: '. . . the process whereby an electron and a positron unite and consequently lose their identity as particles transforming themselves into short gamma rays' (Webster's Third New International 1971, p. 87). In this sense, annihilation means something more than the mere disappearance and end of phenomena: a stage of merging occurs before one thing is created from these two forms. Out of nothing does come something—at least in theoretical physics.

In Buddhist philosophies too, annihilation of the ego is the highest state of being a human can attain. Japanese roboticist Masahiro Mori, theorist of the uncanny valley, writes, 'human beings have self or ego, but machines have none at all. Does this lack cause machines to do crazy, irresponsible things? Not at all. It is people, with their *egos* who are constantly being led by selfish desires to commit unspeakable deeds. The root of man's lack of freedom (insofar as he actually lacks it) is his *egocentrism*. In this sense, the *ego-less* machine leads a less hampered existence' (Mori 1999, p. 49; my emphasis). In the *Buddha and the Robot*, Mori writes of his Buddhist vision of science, technology and robots. In Buddhist philosophies, the relations between different kinds of things are seen as interrelated. 'As I consider questions of this sort, I am reminded of the Buddhist axiom that "nothing has an ego". This means that nothing exists in isolation; everything is linked with everything else' (Mori 1999, p. 28).

Cartesian dualism, which proposed the mind as transcendent and the body as immanent, did capture something about the nature of ontological difference. In rejecting Cartesian dualism, anti-dualist categories have emerged ('cyborgs', 'meshworks', 'actor-networks', and 'assemblages'), but such styles propose a multiplicity without proposing any ontological difference of the different entities. Cartesian dualism has not been resolved—it has been side-stepped into a form of merging. In rejecting the ontological difference that was captured in the theorizing of dualism, a form of the "I/ego" is also threatened: the "I" as a human subject and different from other entities. Robotic scientists and AI theorists bring these issues to the fore in the way they create artificial beings (Helmriech 1998).

EVERYTHING IS CONNECTED

"Everything is connected" is a phrase we hear repeatedly: from chaos theory, when a butterfly flaps its wings, creating havoc elsewhere (Gleick 1994), to globalization (Erikson 2003) that emphasizes global flows of markets, labor, goods, services and capital. Let us consider Marilyn Strathern's (2014) points in relation to this statement:

> Indeed, the more so-called 'bounded' notions of society and culture are held up to criticism, along with the systems and structures that were once their scaffold, the more relations, relationships, the relational, relationality, are evoked as prime movers (of sociality) in themselves. Quite aside from identifying relations in structures, systems of classification, co-variation, and so forth, the concept is equally forcefully applied to any new object of knowledge, emergent configuration, or co-construction, and not only in a passive sense (*everything is connected*), *but in the active sense of the observer making phenomena appear, illuminating them, by the concept.*
>
> (p. 5; my emphasis)

In the 'active sense of making phenomena appear', Strathern highlights the construction of connections between everything. Strathern (2014, p. 10) takes up these points and develops philosopher John Locke's ideas of association when what becomes connected to something else is dependent on the types of associations that are crafted. Locke's theory of associations was also a theme that interested cybernetics pioneer Norbert Weiner. In Weiner's classic text

Cybernetics: or, Control and Communication in the Animal and the Machine,
he outlines a theory of cybernetic systems as organic, mechanical and tied
together by control and communication systems (1961). Wiener draws on
Locke's themes, exploring 'the possibility of assigning a neural mechanism to
Locke's theory of the association of ideas'(Wiener 1961, p. 156). Weiner uses
the example of recognition of the face of a man to explore this issue: 'how
do we recognize the identity of the features of a man, whether we see him
in profile, in three-quarters face, or in full face? (1961, p. 156). Weiner, like
Strathern, was interested in the parts of the person, and how ever-diminishing
parts could still stand in for the whole of the person (Strathern 1988).

The feminist model of the cyborg developed by Donna Haraway needs
to be honored in this history of humans and machines, and though written
over 20 years ago, the cyborg reveals something distinctive when contrasted
with different cultural imaginations of robots. The cyborg is an analytical
device with which to assess the breakdown of organism and machine as
distinctive categories:

> Although the cyborg image originated in space and science fiction to
> refer to forms of life that are part human and part machine, it is by no
> means confined to the world of technology. Rather, cyborg anthropol-
> ogy calls attention more generally to the cultural production of human
> distinctiveness by examining ethnographically the boundaries between
> humans and machines and our vision of the differences that constitute
> those boundaries.

> (Downey, William & Dumit 1995, pp. 264–265)

The cyborg was appropriated by Haraway as a polemical tool to critique social relations, and in this sense, it is similar to the robot. Whereas
the robot as imagined by its creator Čapek expressed the fear of boundary transgressions between human and nonhuman, Haraway's cyborg takes
them as given, and she pushes the boundary transgressions further in her
political work:

> I want to signal three crucial boundary breakdowns that make the fol-
> lowing political-fiction (political-scientific) analysis possible. By the
> late twentieth century in United States scientific culture, the boundary
> between human and animal is thoroughly breached. . . . The second
> distinction is between animal-human (organism) and machine. . . . The
> third distinction is a sub-set of the second: the boundary between physi-
> cal and non-physical is very imprecise for us'.

> (1991, pp. 151–153)

For Haraway, the cosmologies that constitute modernism have been
called into question via new technologies and feminist theorizing. The

cyborg is a symbol of the breach of boundaries and their playful dissolution. The cyborg may have had its heyday in the 1980s, but it is an important analogical (and digital) symbol for theorizing about the robot. The cyborg is an anti-dualistic and anti-essentialist symbolic construct, in the sense that Haraway in her essay critically attacks patriarchy, colonialism, and capitalism, drawing the lines between these positions and social theorizing of lived realities (1991). The robot, too, once served the same purpose as an object critical of modernism and Enlightenment, but Čapek's robots were a nightmare, not an ironical celebration as our latter day cyborg.

Tim Ingold (2012) proposes an alternative framework to Haraways' cyborg of leaky machine-organism configurations, noting the situated aspects of becoming between person and environment that he calls "meshwork", as he explains: 'Together, these entangled lines, of bodily movement and material flow, compose what I have elsewhere called the meshwork, as opposed to the network of connected entities. And this *meshwork* . . . is nothing other than the web of life itself' (p. 435; my emphasis). Ingold is distinguishing himself from the networked model of associations proposed by Latour (2005), but in the meshwork, there is no distinctive agent. In my reading, the meshwork is another kind of network. Whereas the coordinates are set in different configurations, the essentialist aspect of the human actor is still lost. Meshwork echoes the term "enmeshment", a condition of being unable to separate that speaks to being trapped, as in a net—a mesh. These lines of enquiry in anthropological theorizing reduce the human to nothing, speaking instead of a multiplicity made up of many parts.

Latour (2005) and Rabinow (2011) prefer to use the term 'assemblages' to describe these complex multiplicities of humans and nonhumans. Strathern takes up Rabinow's themes, and in *Reading Relations Backwards* she writes:

> Assemblages are composed of preexisting things that, when brought into relations with other preexisting things, open up different capacities not inherent in the original things but only come into existence in the relations established in the assemblage.
>
> (Rabinow cited by Strathern 2014, p. 4)

What does it mean when human and nonhumans are 'assemblages', 'networks' or 'meshworks'? What does it mean when 'preexisting things, when brought into relations with preexisting things, open up different capacities not inherent in the original'? To say that the original is continually emerging as original is an intriguing position. Is creativity really an outcome of endless assemblages of different things? This is what the architectural team at MIT thought when they were building the Frank Gehry building—they called this design framework 'communicative sociality', which accounted for the extensive open-planned design of the building's interior. A year after arriving at MIT, I, along with the entire research group, moved into the Ray and Maria Stata Center in 2004. The building is famous on the MIT campus as it is designed by architect Frank Gehry. The building was unfinished and furiously disliked by the new occupants. One of their architect team explained to me this philosophy of 'communicative sociality':

> When someone is walking through one space and they can see something on a board, then they walk through another space and see something else, they can be creative because they can put together different parts.
> (Personal communication 2004)

The Gehry architect team was creating a version of Rabinow's 'assemblage' in the built environment. The irony was that the Gehry building was built on the old ruins of Building 20, a prefabricated building that was only meant to last a few years but was kept on for nearly 40, until it was demolished to make way for the Gehry building. Building 20 was known for the innovative development of radar in a building that cost a few dozen thousand dollars with no such design philosophy as opposed to the Gehry building, which cost $300 million (Dey 2007).

As in the design of the Gehry building or in relation to Rabinow's concept of the assemblage, we must ask: is any part of the assemblage any different from the other? There is an apt analogy in computing that is a kind of assemblage device: the compiler. Compliers are computer programs used by robotic scientists to bring different mechanical and electrical systems together.

Computer analogies are of primary importance in shaping the idea of a "network", a theory developed by Latour to describe a multiplicity of parts with no central actor. Tim Berners-Lee, founder of the World Wide Web, initially designed a software program called Enquire, which stood for 'Enquire Within upon Everything' (1999, p. 1). Berners-Lee was driven by his interest in connections between different entities while a researcher at CERN in Switzerland.

In the chapter entitled "Tangles, Links and Webs", Berners-Lee explains his interests that foregrounded his research into web systems:

> In an extreme view, the world can be seen only as connections and nothing else. . . . There really is little else to meaning. The structure is everything. There are billions of neurons in our brains, but what are neurons? Just cells. The brain has no knowledge until connections are made between neurons. All that we know, all that we are, comes from the way our neurons are connected.
> (2007, p. 14)

The flat ontological model of interconnected nodes, assemblages, cyborgs, meshworks and actor-networks show resemblances between anthropological theorizing, and robotic and AI models of information systems. As an anthropologist using theoretical models developed in my field to help me explain the ethnographic data I found while at MIT, I found the point of resemblances worthy of note.

I want to assert that there are many similarities in the frameworks developed by anthropological theorists, particularly their focus on the flat ontological and decentralized anti-hierarchical systems that make up human-machine modalities and computer scientists, AI scientists and robot-icists. Robotic scientists and AI researchers at MIT are involved in creating new kinds of artificial entities and using these entities as a way to reflect back on the human.

I will show how the fields of robotic science and AI share an underlying openness about human-nonhuman relations, and this is accompanied by the radical redefinition of the social. In this book I call this mechanical sociality.

MECHANICAL SOCIALITY

Over the last few decades, the meaning of the social as a distinctively human quality has been disputed (Appadurai 1986; Latour 1993; Latour 2001; Latour 2005). The "social" is so thoroughly detached from human sub-jects that it can be found everywhere and focused on the effects of agents (Latour 2001). In robot labs the emergence of the social robot in the early 2000s showed a new way of reflecting on human-robot relations. Artificial intelligence, developed as a sub-field of computing in the 1950s, focused on simulating human intelligence in machines; the human subjectivity was within a computer system, and the body of a machine was irrelevant to these functions. Robots by contrast offered a way to write the social on the body of the machine by focusing on the way that it acted with humans. Robotic scientists called this 'situated learning', 'situated action', or 'behavior-based robotics'. All these different configurations focused on the present aspects of the machine *in situ*. For roboticists, the social is based on a socially interac-tive ritual. If the robot acted in a "social" way and could entice people in an interaction, this was a testament to its success as a social machine. Therefore the social in robotics is located in the interpersonal space between pres-ent actors: it is in the micro-exchange of human and robot. Digital online social networking by contrast carries its social meanings through networks. Whereas digital social action is important, I will confine my discussion of the social to how it is configured by robotic scientists, adjusting and reflect-ing on these practices as they emerged in the making of robots at MIT.

To speak of mechanical sociality is not possible without reference to Dur-kheim, who in different words described traditional society as 'mechanical' and industrial society as 'organic'—he did not refer to sociality, but to soli-darity (1965). These terms are important. For Durkheim, a sociologist and witness to the cultural transformations of modernity, solidarity is the bind and bonds between people, featured as the glue that connected person to person in society (pre-system theories). Durkheim's emphasis on traditional society as a kind of machine and modernity as organic is worthy of some reflection, not least because in postmodernity we seem to be enmeshed in the mechanical networks. This posthumanist turn in anthropological theorizing underscores these theoretical models, and anthropologists like Karen Barad (2003) argue for a 'posthuman performativity' and 'intra-agentive realism', rejecting the dualistic separation between 'observer' and 'observed'. Barad writes that, 'On my agential realist elaboration, phenomena do not merely mark the epistemological inseparability of "observer" and "observed"; rather, phenomena are the ontological inseparability of agentially intra-acting "components."' (2003, p. 815). Barad rejects ontological separateness and the distinctiveness of different agents.

Whereas traditional society was locked in a kind of prefigured structure, modern society was changing, even progressing. Durkheim's nineteenth-century model of the social has been dethroned by contemporary sociologists of science such as Latour, who has resurrected a multiple model of the social, in contrast to Durkheim's peculiarly human one. In a recent paper, 'Gabriel Tarde and the End of the Social', Latour writes that 'actor-network theory', or ANT, is a deliberate attempt to terminate the use of the word 'social' in social theory and to replace it with the word 'association' (Latour 2002, p. 117). Strathern (2014, p. 7) makes a relationships between Locke's theory of association, which she regards as a historical preliminary to Rabinow's description of 'assemblages'. The social as a specifically human feature acted as the foundation for separating out the human and nonhuman parts. Humans were 'socialized', that is turned into particular kinds of persons, in modern societies (Simmel in Joyce 2002, p. 1), or they lived in 'society' envisioned as 'a kind of social interaction lying somewhere between the realms of "the family" and "state"' (Withington cited by Strathern 2014, p. 10).

If anything, robotic scientists were consciously creating kinds of "humans" in the robots they created and put a great emphasis on the 'social', whereas at the same time, social scientists led by sociologists of science such as Latour abandoned the concept of the 'social' (2002). Helmreich's (1998) pioneering study of artificial life at the Sante Fe Institute in the US in the 1990s repeatedly asserts that the researchers at the institute are ignorant of normative models of sex, race, gender and class that are refigured in the computer simulations of artificial life, but in my case, the robotic scientists were conscious of race, class and gender, and none wanted to reproduce these normative stereotypes in the robots they created. The research scientists at MIT avoided racially marking the "skin" of their creations by making sure there was no confusion that it was a machine, keeping all its machined parts on show and avoiding any racially marking colors. The robotic scientists avoided reproducing sexed machines, preferred to keep their machines genderless, and did not speak in class-marked categories of their robots as "servants" or "workers", but companions, friends and children.

Robotic scientists propose the social is an outcome of an interaction between human and robot. In the making of social robots at MIT, these robots were imagined as children. Adults were encouraged to relate to the machine in the role of a parent or caregiver. To facilitate these kinds of exchanges, robots are designed in particular ways—as non-threatening, even cute. Robots as children and humans as parents open up the possibility of new kinds of attachment patterns, which will lead us to explore the science of attachment and ask fundamentally how humans are able to form bonds with other humans and if such attachment patterns can be transferred to machines.

Perhaps the emphasis of robotic scientists to create social robots and to model them as children is a way of highlighting the importance of childhood in contemporary social life. While these robots were funded by the Defense Advanced Research Project Agency (DARPA), an arm of the US military, the research robots were less geared towards their military purposes. Rather insidiously, the findings of these childlike robots are taken back into the military. Military organizations fund research in academic institutions but develop their military devices in their own research centers. The robot Kismet is a case in point. A "cute" and childlike disembodied humanoid head, it was designed by a team of researchers at MIT lead by Cynthia Breazeal and was funded by DARPA and grant funds from the MARS project. As an expressive robot face that can smile, frown, show boredom and sigh, this shows the sometimes contradictory nature of research in the lab and the organizational ethos financially supporting it.

What is the model of the human that underscores anthropological theorizing? It seems fitting that a discipline that evolved out of the study of "human" ("anthropos") should itself be at the forefront of reducing to nothing the human agent. 'Modernity', Latour writes, 'is often defined in terms of humanism, either as a way of saluting the birth of "man" or as a way of announcing his death' (1993, p. 13) For Latour, epistemological problems emerge because 'Modernity . . . overlooks the simultaneous birth of "nonhumanity"—things, or objects, or beasts—and the equally strange beginning of a crossed-out God, relegated to the sidelines' (1993, p. 13). There is a methodological resolution in the above: how to overcome dualisms without reverting to opposite forms of thinking, and therefore returning to dualistic thinking.

In *We've Never Been Modern*, Latour argues that 'Postmodernism is a symptom, not a fresh solution. It lives under the modern Constitution, but it no longer believes in the guarantees of what the Constitution offers' (1993, p. 46). Latour argues that the 'premodern', 'modern' and 'postmodern' are all interrelated categories, built out of each other. The category of the 'modern' emphasizes the purification of entities—nature/culture, human/nonhuman. The category of 'premodern' emphasizes the hybridism of entities, and the 'postmodern' rejects the modern critique 'but it [modernity] is not able to do anything but prolong that critique, though without believing in its foundation' (1993, p. 46). Latour aims to rework the essentialist categories of what constitutes agents: 'Action is simply not the property of humans but of an association of actants . . .' (1999, p. 182). What are the consequences of action when it is not simply a property of humans? Latour's concept of hybrids constitutes the interrelationships between humans and nonhumans. The 'work of purification' is what Latour assigns to the moderns, while actor-network theorists explore 'translation' and the formation of 'hybrid networks' (1993, p. 11). The project for moderns is to purify the hybrids; otherwise the hybrids represent 'the horror that must be avoided at all costs by a ceaseless, even maniacal purification' (Latour 1993, p. 112). What is sacrificed in this approach? Latour's theorizing reformulates social relations between distinctive categories of humans and nonhumans in favor of hybrid networks: 'we have to turn away from an exclusive concern with social relations and weave them into a fabric that includes non-human actants, actants that offer the possibility of holding a society together as a durable whole' (Latour 1991, p. 103).

Ultimately, Latour rejects ontological, essential differences, writing 'I have simply re-established symmetry between the two branches of government, that of things—called science and technology—and that of human beings' (1993, p. 138). In re-establishing symmetry between different kinds of things, the consequential analytical process does two opposing analytical things: it makes things appear radically separate and distinct, detached and chaotic, but on the other hand, it merges them into one.

Is there another route out of Latour's model—is defending human subjectivity always an act of purification? In what follows, I want to show how the making of robots calls into question these positions by focusing on those technologies that comfortably embrace the hybrids, that model and develop the hybrids in new ways. Robots and AI systems show the limits of hybrid systems. Artificial intelligence and social robots try to create new hybrid forms where human-nonhuman attachments are reconfigured.

HOW TO ATTACH A HUMAN TO A MACHINE

In exploring the applied aspects of social robotics and social machines, the links between attachment and detachment will come to the fore.

In *Love and Sex with Robots: The Evolution of Human-robot Relationships* (2009), computer programmer David Levy makes the case for human-robot relationships, but he premises this possibility on observing human-human relationships that presently exist and exhibit a state of detachment. Levy writes about the widespread use of sex-workers and proposes that robots could fill these roles. Levy's case for robots is that human relationships are too messy and complicated; there would be no misunderstanding between a robot programmed to do what the human wanted. Levy further speculates that if the robot becomes too predictable and submissive, you could alter its program so it could demonstrate more resistance to you. Levy's book is an argument for why sex and love with machines is possible—because men and women both seek sexual pleasure in the absence of a full person (with a substitute sex simulator such as a vibrator, or via sex-work). He explains:

> More obvious reasons why the robot experience will be more appealing than visiting a prostitute include the utterly convincing manner in which robots will express affection and other emotions, simply because their emotions will be programmed into them, to be part of them, instead of being make-believe affections acted out by a prostitute with little genuine enthusiasm for the need to convince.
>
> (Levy 2009, p. 206)

This theme of lack of empathy is taken up by autism expert Baron-Cohen (2011), and he argues in *Zero-Degrees of Empathy* that violence, abuse and murder are only possible because individuals lack empathy, namely men. Baron-Cohen proposes that autism is a state of 'zero-degrees' of empathy and is an extreme form of the male condition. Robots come to the rescue in helping to support those children and adults where attachments are broken or disordered, and autism spectrum disorder is a case in point. There is a narrative about human relationships presently seen as confusing, unsatisfying and conducted without empathy and attachment. If humans already relate to each other in these ways and a machine can act as the human might in the same circumstances, then it could become a viable alternative. Robots are proposed to fill these roles.

Robotic scientists at MIT by contrast are tying to cultivate affectionate bonds between humans and robots, but the social is reworked as performative and scripted, a set of acts that are predictable. Robotic scientists did not design their robots to resemble sex-workers as Levy hopes, but instead, robots in labs at MIT (and beyond) are crafted in the image of children. The philosophy of these robotic scientists is to create robots as relational companions to humans. It was among robotic scientists in labs at MIT that I first heard the term "companions" to refer to extending the companion range of "significant otherness" to nonhumans. Donna Haraway took up these themes in *The Companion Species Manifesto: Dogs, People, and Significant Otherness* (2003), proposing a lifeworld of natureculture mixtures of humans and dogs. Underscoring such positions is a model of sociality for anthropologists interested in multispecies ethnographies (Kirksey & Helmreich 2010) and otherness (Haraway 2003). For robotic scientists, it was also about extending the relational possibilities to robotic machines.

Robots are imagined to help fill the gaps in human social relations and are imagined to become friends and companions to a growing elderly population (Robertson 2007), therapeutically support children with autism and to be sexual companions (Levy 2009, Robertson 2010). In *Alone Together: Why We Expect More from Technology and Less from Each Other,* sociologist of technology Sherry Turkle (2011) warns that human social relationships are under threat by the artificial, a topic she has researched extensively since the publication of *The Second Self* (1984), where Turkle explored how children were attaching to newly computerized technologies.

Why are robotic machines emerging to help human relationships? While the social may not be exclusively human, as seen in other living creatures living in social collectives (Latour 2001; Haraway 1991; Enfield & Levinson 2006), the extension of the social to machines is unique to the contemporary age. The corporate message from Silicon Valley is the "social"—social computing, social networking, social machines, and alongside this, social robots. At the interpersonal level, the social is the mutual dialogical space between one person and another (Buber 1937; Stawarska 2009). The social is made up of gestures, vocalizations, speech, behavior, shared attention, cognition and affective exchange (Enfield & Levinson 2006). If there is an impairment in reading these cues, it can result in difficulty for children and adults. Children and adults with autism have difficulty reading the cues of another person and making sense of their behaviors (Baron-Cohen 1995). As autism is a social interaction difficulty, it may come as no surprise that robotic scientists at MIT began to imagine a robotic machine as a kind of person with autism, an entity that lacks the capacity to read social cues and respond appropriately (Scasselati 2001). We find that analogies between people with disabilities and machines recur as aspects of the making of robots.

In technology circles, the social is valorized, given a meaning and a life of its own. We can attribute some of this to Marx, who in *Capital* wrote: '. . . the relations connecting the labour of one individual with that of the rest appear, not as direct social relations between individuals at work, but as what they really are, material relations between persons and social relations between things' (1974, p. 78). Particular kinds of persons have been identified by robotic scientists as potentially able to benefit from their technologies: older populations (Robertson 2007), children with autism (Dautenhahn & Werry 2004)and (perhaps one day when the technology is sufficient), adults, some who have difficulty forming emotional attachments and desire a sexual relationship (Turkle 2011; Levy 2009). These different populations suffer from severe attachment issues.

Robots are created to help humans in all these areas of their existence. I call all these different types of attachment difficulties "attachment wounds". The machines are imagined to save us from modern attachment wounds. The machines can act in place of another person—as a lover, a friend or as a therapeutic agent.

Durkheim's concept of "anomy" is described as a state of detachment from lived existence in modernity (Durkheim 1952). Anomy referenced a state of despair felt at the level of the individual in society. While Durkheim described attachment and detachment through the discourse of solidarity and anomy respectively, it was not until the early twentieth century that theorists began to think about how attachments are made between humans.

Robotic scientists propose that humans can assist and help the robot to develop, which then returns us to the primary questions: How do humans actually form bonds with one another? How do humans make one another? What happens when there is disruption in the bond?

In his seminal work *Attachment and Loss* (1981), John Bowlby outlines his attachment theory, emphasizing the importance of a loving and stable bond for children to grow in prosperity. Bowlby identified three important stages in the attachment process. The child would exhibit neurotic disruption expressed as protest, despair and detachment if the attachment process was disrupted by the death of a parent, war or other crises that may disturb the secure relationships of the child with a primary caregiver or givers. The new science of attachment that Bowlby created was an outcome of his work for the World Health Organization (WHO) in the 1950s, where he prepared a report on the adjustment of children's mental health, writing 'What is believed to be essential for mental health is that the infant and young child should experience a warm, intimate and continuous relationship with his mother (or permanent mother-substitute) in which both find satisfaction and enjoyment' (1981, p. 12). It was not only Marxist feminists who resisted the motherly attachment logic of Bowlby; psychoanalyst Melanie Klein (mentor of Bowlby) wrote 'Dr. Bowlby, we are not concerned with reality, we are concerned only with the fantasy' (Kagan cited in van der Horst 2011, p. 21). Bowlby's work was focused on the mother-child relationship and separations between them, as well as the mother's treatment of the child (unconscious attitudes), and illness and death in the family (Bowlby 1981, pp. 21–22). The importance of a loving attachment for the development of a child was paramount. For Bowlby, the 'development processes' of attachment were vital to making humans distinctive. He wrote: 'The truth is that the least-studied phase of human development remains the phase during which a child is acquiring all that makes him most distinctively human' (Bowlby 1981, p. 423).

Sigmund Freud was arguably the first to take seriously childhood experiences as important in shaping the lifeworlds of adults, and for connecting the patterns of childhood experience with adult neurosis (Freud 2003; Bowlby 1981, p. 424). In *Leonardo da Vinci and A Memory of His Childhood*, published in 1910, Freud proposed that da Vinci's subject matter in his art could be traced back to his childhood. Freud used biographical accounts of da Vinci, alongside analysis of his paintings, such as *The Virgin and Child with St. Anne*. The virgin with St. Anne, Freud believed, was really da Vinci's own relation to his early experiences, having two mothers, a birth mother (Caterina, a peasant woman) and a stepmother. Da Vinci's father married another woman Albiera, and da Vinci was conceived out of wedlock (Freud 2003).

The robotic scientists also channeled their psychical-physical sufferings into the robots they created, and their machines mirrored parts of themselves. Robots were modeled on the unconscious psychic sufferings of their makers, as physical and social limitations and models of post-traumatic stress disorder were imported into the machines. While Freud arguably is the first to acknowledge the importance of childhood in the development of adult neurosis, he famously discounted his patients' reports of sexual abuse as mere fantasies (Hacking 1991, p. 267). Freud was more interested in proving his Oedipus theory and using material given by his patients to illustrate his theory that children want to sexually possess the opposite gender parent. This is illustrated in his analysis in *Screen Memories*, where the life history of his patient is forced into his Oedipal frame (2003). It was up to later psychoanalysts such as Klein and Bowlby to develop perspectives on childhood experience and later adult neurosis, but Bowlby moved on from the focus of Klein's 'object-relations'.

For Bowlby, love between parent and child was essential, and he challenged the widely held belief that the most important aspect of the parent-child relationship was subsistence. He drew on the work of Harry Harlow, who conducted experiments on rhesus monkeys and found that clinging to something soft was more important for the monkeys than food (Kagan 2011, p. xiii). These awfully cruel experiments on primates showed that younger primates required comfort as well as food. Bowlby's theory of attachment brought together multiple strands from psychoanalysis and experimental psychology, ethnology, primatology and the influence of environment on biology (1981).

Attachment ideas were enthusiastically taken up by Bowlby's colleague Mary Ainsworth, who conducted studies on parent-child relationships in Uganda, the US and England. In Uganda (1954–1955), she spent a period observing 28 Ganda infants, and from here she proposed three patterns of attachment: secure attachment, insecure attachment, and non-attachment (Ainsworth 1967). Ainsworth and colleagues (1978) went on to explore the important of attachment in the *Strange Situation Experiment*, and argued these attachment models shape the child's psychology and ability to form relationships with others.

In the 1940s another field of childhood psychiatry emerged, studying a pattern of detachment in children that would be termed autism (Kanner 1943). Autism, "first" identified by psychiatrist Leo Kanner in children, was marked by an absence of relating, repetitive behaviors, and speech, language and communication difficulties (Kanner 1943). Kanner wrote of children with autism:

> The Children's *relation to people* is altogether different. Every one of the children, upon entering the office, immediately went after blocks, toys or other objects, without paying the least attention to the persons present. It would be wrong to say that they were unaware of the presence of persons. But the people, so long as they left the child alone figured in about the same manner as did the bookshelf or the filing cabinet.
>
> (1943, p. 242)

In Switzerland in 1944, another psychiatrist, Hans Asperger, also published reports of children he described as exhibiting severe social-interaction difficulties (Asperger 1991). Prominent autism expert Uta Frith had this to say about Kanner and Asperger: 'By a remarkable coincidence, Asperger and Kanner independently described exactly the same type of disturbed child to whom nobody had paid much attention before and both use the label autistic (Frith 1991, p. 6). Autism and attachment theories arose almost simultaneously from the 1940s. Autism was marked in the biomedical framework as an outcome of biology and genes, and more recently neuroscience rather than "nurture" (Baron-Cohen 2005). Bruno Bettleheim is controversial in this respect, and he argued that autism was a survival response to cold and detached mothering sometimes coined 'refrigerator parenting' (1967). Bettleheim's controversial position is rejected by autism experts, who instead focus on the neurodevelopmental aspects of autism (Baron-Cohen 2011, Frith 2008).

Attachment theory is a description of how humans (caregivers) make other humans (infants), and turn these children into particular kinds of persons. Bowlby, Ainsworth and Bettleheim propose the social is the mutual and dialogical interaction between parent/caregiver/mother and infant/child. Melanie Klein proposed an alternative theory—'object-relations', that was focused not on the human-human interaction, but how feelings, thoughts and experiences about those relationships were projected onto things (Gomez 1997). Klein's thing-focused view of social relations was arguably extended by Appadurai in his famously titled edited volume *The Social Life of Things* (1986) that focused on commodities, and extended this view that "commodities" (nonhumans) can have a social life.

Robotic science extends and reframes the social through its activities. By designing robots like children, robotic scientists attempt to cultivate affective exchanges between adults and machines. The bond is no longer exclusively human. Computer scientist Josef Weizenbaum (1984) warned against human attachments to machines after he discovered his computer program ELIZA (the first chatbot) became a close confident of his colleagues, causing Weizenbaum to begin to worry about the future of humanity. For robotic scientists, the social is located in micro behavioral exchanges between human and robot. The narrative of attachment and detachment plays out in multiple ways in anthropological theorizing, and in robotic and AI research.

In what follows, I hope to unfold these issues addressed above in different ways.

In chapter 1 I explore the origin of the robot as a cultural outcome of the cultural milieu of the 1920s. Drawing on themes in the play and connecting these theme to actual events occurring in the early twentieth century, I show how the robot was a critical response to what Čapek believed was a obsession with labor and production by right and left political philosophies. I explore the role of revolution and the fear that humans were losing their individuality.

In chapter 2 I take the robot over from its fictional and political imaginings and bring it into the fold of artificial intelligence, a technological field focused on simulating human intelligence in machines. I continue the theme of distorted attachments by showing how much of the efforts of AI were devoted to developing war machines. I focus on the personal biography of Alan Turing and show how his biographical story was connected with his theorizing of thinking machines. If AI focused on the development of Cartesian rationalistic disembodied minds, the rise of embodied robotics and behavior-based robotics were anti-Cartesian in orientation, rejecting the notion of a centralizing consciousness or "mind". Instead, these researchers proposed that consciousness resides in bodies, behaviors and movements that allow others to interpret them.

In chapter 3 I then explore the philosophy around social robots and social machines. These robots were developed in labs at MIT and beyond, and signaled new ways of reflecting on what it means to be social and how sociality between humans and machines can develop.

In chapter 4 I take these themes of the social and asocial by exploring the kinds of gendered persons that are involved in making robots and AI systems. These types of persons, stereotypically labeled "nerds" or "geeks", show the paradoxes at play with asocial researcher scientists developing social systems. The asymmetries and symmetries between humans and machines offset each other frequently.

In chapter 5 I explore how the robot creation represents a figure of human suffering and breakdown, and show how robotic scientists used their machines as an unconscious dialogue sounding board for their own existential anxieties and difficulties. Robot and robot scientist seemed to mirror each other in unusual ways.

In chapter 6 I explore the role of fantasy and the Real in the making of robots. These robots were premised on a philosophy of the Real—they were robots designed to act in the Real world, but, as my ethnography shows, there were ongoing tensions between the robot and the Real world, the two reflecting on each other in unique ways.

NOTE

1. In January 1985, less than one year after released in 1984, *Terminator* had grossed $38,371,200 in the US alone.

BIBLIOGRAPHY

ABI Research 2011, *Military robot makers to exceed $8 billion in 2016.* Available from: <https://www.abiresearch.com/press/military-robot-markets-to-exceed-8-billion-in-2016>. [10 November 2014].

Ainsworth, M 1967, *Infancy in Uganda*, Johns Hopkins Press, Baltimore.

Ainsworth, M, Blehar, MC, Waters, E & Wall, S 1978, *Patterns of attachment: a psychological study of the strange situation*, Lawrence Erlabaum Associates, New York.

Appadurai, A 1986, 'Introduction', in *The social life of things: commodities in cultural perspective*, ed. A Appadurai, Cambridge University Press, Cambridge, pp. 3–63.

Asperger, H 1991, *'Autistic psychopathy' in childhood*, trans. U Frith., in *Autism and Asperger syndrome*, ed. U Frith, Cambridge University Press, Cambridge, pp 37–92.

Barad, K 2003, 'Posthumanist performativity: toward an understanding of how matter comes to matter', *Signs: Journal of Women in Culture and Society*, vol. 28, no. 3, pp. 801–831.

Baron-Cohen, S 1995, *Mindblindness*, MIT Press, Cambridge, Mass.

Baron-Cohen, S 2003, *The essential difference: men, women, and the extreme male brain*, Basic Books, New York.

Baron-Cohen, S 2011, *Zero-degrees of empathy: a new understanding of cruelty and kindness*, Allen Lane, London.

Berners-Lee, T 1999, *Weaving the web: the past, present and future of the world wide web by its inventor*, Orion Business Books, London.

Bettelheim, B 1967, *The empty fortress; infantile autism and the birth of the self*, Free Press, New York.

Bowlby, J 1981, *Attachment and loss*, vol. 1, Harmondsworth, Middlesex, London.

Bowlby, J 1988, *Attachment and loss*, vol. 2, Pimlico, London.

Bowlby, J 1998, *Attachment and loss*, vol. 3, Pimlico, London.

Buber, M 1937, *I and thou*, trans. RG Smith, T. & T. Clark, Edinburgh.

Buchli, V 1997, 'Khrushchev, modernism, and the fight against "petit-bourgeois" consciousness in the Soviet home', *Journal of Design History*, vol. 10, no. 2, pp. 161–176.

Buchli, V & Lucas, G 2001, *Archaeologies of the contemporary past*, Routledge, London.

Čapek, K 2004, *R.U.R. (Rossum's universal robots)*, Penguin Classics, New York.

Csordas, T 1999, 'The body's career in anthropology', in *Anthropological theory today*, ed. H Moore, Polity Press, London, pp. 172–205.

Dautenhahn, K, Nehaniv, CL, Walters, ML, Robins, B, Kose-Bagci, H, Assif Mirza, N & Blow, M 2009, 'KASPAR—a minimally expressive humanoid robot for human-robot interaction research', *Applied Bionics and Biomechanics*, vol. 6, no. 3-4, pp. 369–397.

Dautenhahn, K & Werry, I 2004, 'Towards interactive robots in autism therapy Background, motivation and challenges', *Pragmatics & Cognition*, vol. 12, no. 1, pp. 1–35.

Dey, A 2007, 'MIT sues Gehry firm over Stata problems', *The Tech* 9 November. Available from: <http://tech.mit.edu/V127/N53/lawsuit.html>. [10 November 2014].

Downey, GL, Dumit, J & Williams, S 1995, 'Cyborg anthropology', *Cultural Anthropology*, vol. 10, no. 2, pp. 264–269.

Durkheim, E 1952, *Suicide: a study in sociology*, trans. JA Spaulding & G Simpson, ed. G Simpson, Routledge and Kegan Paul Ltd., London.

Durkheim, E 1965, *The division of labor in society*, trans. G Simpson, The Free Press, New York.

Enfield, NJ & Levinson, SC 2006, 'Introduction', in *Roots of human sociality: culture, cognition and interaction*, eds. NJ Enfield and SC Levinson, Berg 3PL, New York.

Erikson, TH 2003, 'Introduction', in *Globalisation: studies in anthropology*, ed. TH Erikson, Pluto, London, pp. 1–17.

Freud, S 2003, 'Leonardo da Vinci and a memory of his childhood', in *The Uncanny*, Penguin Classics, London.

Freud, S 2003, 'Screen Memories', in *The Uncanny*, Penguin Classics, London.

Freud, S 2003, 'The Uncanny', in *The Uncanny*, Penguin Classics, London.

Frith, U 1991, 'Asperger and his syndrome', in *Autism and Asperger syndrome*, ed. U Frith, Cambridge University Press, Cambridge, pp. 1–36 .

Frith, U 2008, *Autism: a very short introduction*, Oxford University Press, Oxford.

Gilbert-Rolfe, J & Gehry, F 2002, *Frank Gehry: the city and music*, Routledge, New York.

Gleick, J 1994, *Chaos: making a new science*, Abacus, London.

Gomez, L 1997, *An Introduction to object relations*, Free Association Books, London.

Graham, E 2002, *Representations of the post/human: monsters, aliens and others in popular culture*, Manchester University Press, Manchester.

Hacking, I 1991, 'The making and moulding of child abuse', *Critical Inquiry*, vol. 17, no. 2, pp. 253–288.

Haraway, D 1991, *Simians, cyborgs, and women: the reinvention of nature*, Free Association Books, London.

Haraway, DJ 2003, *The companion species manifesto: dogs, people, and significant otherness*, Prickly Paradigm Press, Chicago.

Helmreich, S 1998, *Silicon second nature: culturing artificial life in a digital world*, University of California Press, Berkeley.

Hicks, H 2002, 'Striking cyborgs: reworking the "human"', in *Reload: rethinking women and cyber culture*, eds. M Flannagan & A Booth, MIT Press, Mass., pp. 85–106.

Ingold, T 2012, 'Toward an ecology of materials', *Annual Review of Anthropology*, vol. 41, pp. 427–442.

Joyce, P 2002, 'Introduction', in *The social in question: new bearings in history and the social sciences*, ed. P Joyce, Routledge, London, pp. 1–18.

Kagan, J 2011 'Foreword', in *John Bowlby*, 1st edn. ed. F Van der Horst, Wiley-Blackwell, Chichester, West Sussex, UK.

Kanner, L 1943, 'Autistic disturbances of affective contact', Nervous Child, vol. 2, pp. 217–250.

Kirksey, SE & Helmreich, S 2010, 'The emergence of multispecies ethnography', *Cultural Anthropology,* vol. 25, no. 4, pp. 545–576.

Latour, B 1991 'Technology is society made durable', in *A sociology of monsters: essays on power, technology and domination,* ed. J Law, Routledge, New York, pp. 103–131.

Latour, B 1993, *We have never been modern*, trans. C Porter, Harvester Wheatsheaf, New York.

Latour, B 1999, *Pandora's hope: essays on the reality of science studies*, Harvard University Press, Cambridge, Mass.

Latour, B 2002, 'Gabriel Tarde and the end of the social', in *The social in question: new bearings in history and the social sciences*, ed. P Joyce, Routledge, London, pp. 117–132.

Latour, B 2005, *Reassembling the social: an introduction to actor-network-theory*, Oxford University Press, Oxford.

Levy, D 2009, *Love+sex with robots: the evolution of human-robot relationships*, Duckworth, London.

Marx, K 1974, *Capital*, Lawrence & Wishart, London.

MIT Humanoid Robotics Group n.d., *Frequently asked questions*. Available from: <http://www.ai.mit.edu/projects/humanoid-robotics-group/cog/faq.html>. [12 November 2014].

Mol, A 2002, *The body multiple*, Duke University Press, Durham.

Mori, M 1999, *The Buddha in the robot: a robot's engineers thoughts on science and religion*, Kosei Publishing Co., Tokyo.

Picard, R 1997, *Affective computing*, The MIT Press, Cambridge, Mass.

QS World Rankings 2014, *QS World University Rankings, 2014/15*. Available from: <http://www.topuniversities.com/university-rankings/world-university-ran kings/2014#sorting=rank+region=+country=+faculty=+stars=false+search=>. [12 November 2014].

QS World University Rankings 2013, available from: <http://www.topuniversities. com/university-rankings/world-university-rankings/2013#sorting=rank+region= +country=+faculty=+stars=false+search=>, accessed 22 July 2014.

Rabinow, P 2011, *The accompaniment: assembling the contemporary*, Chicago University Press, Chicago.

Reilly, K 2011, *Automata and mimesis on the stage of theatre history*, Palgrave Macmillan, Basingstoke.

Robertson, J 2007, 'Robo sapiens japanicus: humanoid robots and the posthuman family', *Critical Asian Studies*, vol. 29, no. 2, p. 369–398.

Robertson, J 2010, 'Gendering humanoid robots: robo-sexism in Japan', *Body & Society*, vol. 16, pp. 1–36.

Scassellati, B 2001, *Foundation for a theory of mind for a humanoid robot*, MIT Department of Electrical Engineering and Computer Science, Cambridge, Mass.

Shelley, M 1969, *Frankenstein*, Oxford University Press, Oxford.

Stawarska, B 2009, *Between you and I: dialogical phenomenology*, Ohio University Press, *Athens*, Ohio.

Strathern, M 1988, *The gender of the gift: problems with women and problems with society in Melanesia*, California University Press, Berkley.

Strathern, M 2014, 'Reading relations backwards', *JRAI*, vol. 20, no. 1, p. 3–19.

Suchman, L 2006, *Human and machine reconfigurations: plans and situated actions*, Cambridge University Press, Cambridge.

Turkle, S 1984, *The second self computers and the human spirit*, Granada, London.

Turkle, S 2011, *Alone together: why we expect more from technology and less from each other*, Basic Books, New York.

Van der Horst, F 2011, *John Bowlby*, 1st edn. Wiley-Blackwell, Chichester, West Sussex, UK.

Webster's Third New International 1971, ed. P Babcock, G. & C. Merriam Company, Springfield, Mass.

Weiner, N 1961, *Cybernetics: or control and communication in animal and the machine*, MIT Press, Cambridge, Mass.

Weizenbaum, J 1984, *Computer power and human reason: from judgment to calculation*, Penguin, Harmondsworth.

Films Cited

AI: Artificial Intelligence 2001, dir. Stanley Kubrick & Steven Spielberg.
2001: A Space Odyssey 1968, dir. Stanley Kubrick.
The Day the Earth Stood Still 1951, dir. Robert Wise.
The Matrix 1999, dir. The Wachowskis.
The Matrix Reloaded 2003, dir. The Wachowskis.
The Matrix Revolutions 2003, dir. The Wachowskis.
The Terminator 1984, dir. James Cameron.
Terminator 2: Judgment Day 1991, dir. James Cameron.
Terminator 3: Rise of the Machines 2003, dir. Jonathan Mostow.
Terminator Salvation 2004, dir. McG.
Terminator: Genesis 2015 (forthcoming), dir. Alan Taylor.

1 Revolutionary Robots

Robots are objects of modernity created to reflect on what it means to be human. The robot is a literary creation of Karel Čapek and is first seen in the play *R.U.R.* (Rossum's Universal Robots) in 1921. In *R.U.R.*, the robot is characterized as a human-like entity built to labor and without any other purpose except "to work". This play on the meaning "to work" is critical as it reminds us of Čapek's intention in creating the robot character—to take the worker one step further. The robot character is human-nonhuman, occupying tense spaces in conceptual imaginings (Jiménez and Willerslev 2007). For now we want to locate the robot in its cultural context, the political and artistic vibrancy of the 1920s. It is by locating the robot in this period, when vibrant debates dominated public discourse on the nature of the political organization of society, that robots as revolutionaries become all too apparent. In what follows I explore how the robot play was inspired by themes of work and the radical politics of the early twentieth century in the US, Europe and the new socialist economies in Russia. Čapek invented the concept of the robot as a deliberate device that expressed his horror at the age of mass mechanization and its effects on humanity. Soviet communism, Marxism, capitalism, Fordism and Taylorism contributed to the context in which the robot was first imagined.

THE ROBOT'S GENESIS

Karel Čapek was born on 9 January 1890 in the Habsburg Empire, which became a sovereign state of Czechoslovakia in 1918 and then the Czech Republic after the fall of the Berlin Wall in 1993 that brought about the separation the Czech Republic from Slovakia. Čapek was a playwright and journalist and, as a journalist, he wrote for *Lidove Noviny* (The People's Paper); in 1921, he and his brother Josef became its editors. Before going on to unpick the themes in *R.U.R.*, Čapek's other significant plays included *The War with the Newts* (1937), which is about human-like salamanders taking over the world, and *The Insect Play* (1921), a play he wrote with his brother, Josef. *The Insect Play* is a commentary on Czechoslovakian politics after the horrors of World War I—and most of the characters are nonhuman.

Čapek used salamanders, insects and robots as nonhuman devices in his plays to critically reflect on human life, politics and existence, yet it was the robot that catapulted Karel Čapek to international (and historical) notoriety. The naming of the Robot (he capitalized the R, but I won't unless referring to the play's text), belongs, not to Karel, but to his brother, Josef Čapek, who thought of the term "robot" to describe the artificial beings in his play. In Russian, Ukrainian and other East Slavonic languages, the term "robota" is a word for work but has a specific meaning in Czech referring to the surplus labor of peasants in the feudal economy. It was this latter meaning that inspired Josef to invent the term "robot". Robota (labor debt paid to lords) encompassed many aspects of the labor of peasants, most notably their obligation (Wright 1966). The theme of labor or work is key to its meaning. One might say the robot is another way of saying "an obligated worker". The term "robot" entered the English language after April 1923 when *R.U.R.* was performed by the Reandean Company at St. Martin's Theatre in London. The robot was used initially by Čapek to describe a soulless artificial being who is created to work. What do I mean by soulless? The robots in the play have no independent interiority, feeling, thought or desire and act on commands and instructions from their human masters.

Let us explore some of these themes. The robot rests on the idea that the grand ideal is to build artificial workers to maximize production and free-up human labor from mindless tasks instead created even more problems in society. The substitution of humans by robots led, Čapek imagined, to a loss of human purpose. This theme was a definite reflection on the changing work practices transforming Europe, North America and Russia, which machines and humans tied together in production. Here is my synopsis of the play:

> The play is set on an Island where robots are mass-produced. The factory is run by the general director, Domin. One day, Helena Glory comes to the Island to liberate the robots. Glory believes the robots should have the same rights as humans. In a twist, she is married to Domin. As the years pass, robots are now even more prolific in the world than before. In secret, Helena has persuaded the scientist Hal to change the formula for some of the robots. The robots, with the new formula, can feel and love. These robots lead a rebellion against the humans. As the world comes to an end, the members of the factory reflect on the meaning of life, but all are slaughtered, except for one man, Alquist. Alquist is only allowed to live because he is a worker, too. In another twist, the formula to create the robots is destroyed, and the robots cannot survive if they cannot find a way to produce. It is left to Alquist to try to find the formula, but he fails and, without the formula for robot life, the last surviving human, namely Alquist, and the robots will eventually

perish. Two of the feeling robots, Helana and Primus, choose love over violence.

R.U.R. was very successful throughout Europe, Russia and America in the 1920s. The play premiered at the Garrick Theatre, New York City in 1922 and ran for 184 performances. Čapek's robots captured the imagination of the time and, after the first performance in New York in 1922, critics wrote 'The most brilliant satire on our mechanical age; the grimiest yet subtlest arraignment of this strange, mad thing we call industrial society of today' (cited in Reichardt 1978, p. 36). The success of the term "robot" also replaced "automaton" as the term to characterize working beings. What was perhaps most significant about the arrival of the robot was that it captured a spirit of the age. Čapek created the robot character as a way to comment on what he thought was the contemporary preoccupation with labor and production.

The robot is a creature that is an outcome of separations, transitions and mergers. I identify two important boundary transgressions the robot makes in R.U.R. The first is that the robot is humanlike but is without agency and subjectivity, like a thing. Second, the robot is able to become humanlike by developing subjectivity: feeling and consciousness. The robot is paradoxical: human and nonhuman at the same time. It is an intermediary somewhere between human and nonhuman, a human envisaged as not quite human. In other words, the robot is an intermediary between the human and nonhuman via the role of labor. There is a final theme that could be quite confusing to a reader, that two robots may be able to repopulate the world again because they have found love. The theme of love is an important theme in robot narratives, as is robot destruction. Robot agents in the play become aware of their position in the social order and develop consciousness. The play marks the end of the human, and in violence this new world ruled by robots is created. The robots that choose violence will perish. The robots that choose love will prosper.

THE ROBOT BECOMES A MACHINE

It is easy to see Čapek's robots as a left-wing critique of capitalism, but his views are more complex. Čapek was part of a "radical center" and was opposed to the movements of the right, which he called 'the cold bourgeoisie' and left, which he called 'the revolutionary fire' (1924, p. 3). Čapek's own work critiqued these cultural processes, and his robot provided a means to reflect on the issues of the day. Haman and Trensky point to Čapek's work as a way of reflecting on the politics of the day:

> [Čapek's art] . . . reflects better than anything else Čapek's constant preoccupation with general problems, his weighing and judging of the world and society. There is a pronounced moral standpoint behind this aspect of his art. It is not accidental that we constantly encounter motifs

of judiciary proceedings in his work, for they reflect the author's moral attitude toward life; he presents to his audience the conflicts of this world while passing judgment on them.

(1967, p. 175)

There are many ambiguous themes in the play; one was in relation to the theme of work. A dominant discussion in the 1920s rested on the mass mechanization of commodity production, which rendered the laborer as another 'cog' in the process, just like a mechanism in the machine. The mass labor movements of the time would have also given the author of the play some cause for concern. Čapek's is considered a pragmatist, marginally philosophical outlook, 'which placed a great emphasis upon the relativity of human knowledge and attitudes, and therefore also on the relativity of the so-called great truths, religious and ideological' (Klíma 2004, p. x). Čapek's plays were polemics against the dominant philosophies of the time.

In *R.U.R.* there are several conflicting views on the period. Čapek used each character in his play to exemplify a certain outlook:

> The general manager in the play, Domin, proves that technical progress emancipates man from hard manual labour, and he is quite right. The Tolstoyan Alquist, on the other hand, believes that technical progress demoralises him, and I think he is right too. Bushman thinks that industrialisation is capable of supplying modern needs, and he is right. Helena is instinctively afraid of all this inhuman machinery, and she is profoundly right. Finally, the Robots themselves revolt against all these idealists and, as it appears, they are right, too.
>
> (Čapek cited in Reichardt 1978, p. 40)

Despite the radical right and left philosophies that helped furnish the imaginary context of the robot, Čapek rallied against these philosophies in his own polemical writings. Čapek's position is reflected in the meanings he attaches to the robot. In Russia, the Leninist revolution had occurred less than four years earlier in 1917 while, in the US, large-scale economic expansion was pushing the capitalist system to new heights of economic development. The fate and fortunes of Europe, the US and Russia, left and right, was realized through the application of science and technological reasoning to economic production. Čapek was critical and hostile to these processes and saw the dizzying worship of technological production as suffocating and horrifying. The themes of dehumanization and the impact of the machine had a significant impact on the art of the period (see Wosk 1986 for an interesting overview of the art of the period).

There was little consensus in the politics of the period and, while many condemned the advance of socialism and communism, others, such as the Dada group in France, turned into radical supporters. The ideas of the Dada movement were notably opaque. Raoul Hausman wrote in *Der Dada3* in

April 1920 that, 'Dada is the full absence of what is called Geist. Why have Geist in a world that runs on mechanically' (cited in Benson 1987, p. 46). The machine was worshipped as an antidote to the confusion, chaos and disconnection that modernity delivered to most people's experience. The machine also became a vehicle into a different kind of world. The influence of politics, particularly from the left, influenced all notable artists of the period, from the Russian Constructivists, to Czech Artificialism, to the Italian Futurists (see Benson 1987; Wosk 1986; Vleck 1990).

In Russia, the impact of the constructivists, led by artists such as Alexander Rodchenko and El Lissitzky, showed the importance of modernism. Their work is angular and structured and is a tribute to efficiency and economy (Lodder 2005). The Czech avant-garde was increasing in prominence in the 1920s. Devêtsil, a Czech group of artists, was associated with Karel Teige and had a leftist orientation (Smejkal 1990). Historian of art, Christopher Green, ties the machine to modernism in the 1920s (Green 2006). The body was incorporated into this machine modernist imagery. In the interwar years, the Mensendieck System was influential in Europe and the US. The system was developed by a Dutch-American physician, Bess Mensendieck, who described the body as a machine. Even the activities of housework were included in this rubric (Wilk 2006, pp. 261–262).

Despite the widespread embrace of mechanical metaphors by many artists of the period, Čapek's robots were deliberately non-machine. In the prologue of *R.U.R.*, the character Helena is introduced to a robot, which she cannot tell is a robot. 'Were you born here?' asks Helena to the robot Sulla. Sulla replies, 'I was made here' (Čapek 2004, p. 10). The robots are made of biological material and possess a human-like appearance. Here, Domin explains to Helena Glory how the robots are made:

> The mixing vats for the batter. In each one we mix enough batter to make a thousand Robots at a time. Then there are the vats for the liver, brains, etcetera. Then you'll see the bone factory, and after that I'll show you the spinning mill. . . . The spinning mill for nerves. The spinning mill for veins. The spinning mill for miles and miles of digestive tract are made at once. Then there's the assembly plant, where all of this is put together, you know, like automobiles. Each worker is responsible for affixing one part, and then it automatically moves on to a second worker, then to a third, and so on. It's a most fascinating spectacle. Next comes the drying kiln and the stock room, where the brand new products are put to work.
> (Čapek 2004, p. 13)

Čapek's robots were biologically human and made of flesh and blood but assembled on a mechanical production line and produced with a formula developed by scientists, managers and engineers. Photographs from the first performances of *R.U.R.* in Prague (1921), New York (1922), London (1923) and Paris (1924) all show the robots as human-like. It was the artistic renditions of the play by other artists at the time that gave the robot its metallic form, and it was Čapek who reacted strongly to these new artistic representations. Lamenting after the London theatrical premiere that the purpose of his play was not to reflect on machines but people, he stated: 'I wasn't concerned about robots, but about people' (Čapek 2004, p. xvii). By the 1930s, almost 10 years after the first performance of *R.U.R.*, the metallic robot has become consolidated in the robot's imagery—leading Čapek to distance himself even more from his creation:

> It is with horror, frankly, that he rejects all responsibility for the idea that metal contraptions could ever replace human beings, and that by means of wires they could awaken something like life, love, or rebellion. He would deem this dark prospect to be either an overestimation of machines, or a grave offence against life.
>
> (Čapek wrote this in the third person in 1935)

Čapek could not control how the public viewed, presented or interpreted his robotic entities, despite his protest that he had been misunderstood. It was perhaps inevitable that the robot turned from human being to machine considering the political climate of the period.

Did the robot become mechanical because of the politics of the age? In one way, the robot's turn to machine distanced it from the human. It indicated the malleable status of humanity at the time. It showed the importance of the machine in culturally reflecting human subjectivity. It was certain from Čapek's references in the play, along with his criticism of the mechanization of his creature, that his intention was to examine humans and not machines. But to what extent does the play open up those possibilities? It does this by contrasting humans to robots and machines. Čapek created an in-between object of ambiguous status. Is the robot a human with nonhuman qualities, or a nonhuman with human qualities or neither of these but *other*? Čapek's robot opens up all these possibilities at the same time. The robot is multiple!

In the US, Charlie Chaplin parodied the factory system in his 1930s classic *Modern Times*, while in Germany, the Bauhaus group attempted to rail against the mass production of objects with a respect for craft-based techniques. As Corn and Horrigan explain, 'Robots symbolize the rise of the machine and its replacement of workers; the alienation of humans from their work and from society in general; and the loss of control over the future. . . . But by the end of the 1930s, basically benign, humanoid robots had begun to make appearances at world's fairs and industrial expositions' (1984, p. 74). If there was uncertainty about being human during the 1920s and 1930s, there was also ambiguity about the purpose and nature of machines—machines were cast as dehumanizing and liberating. German film director Fritz Lang released *Metropolis* in 1927 that alluded to a machine-like entity that would eventually replace human laborers. *Metropolis* carries a strong political theme (the workers rebel against their employers) and humans are crafted from a machine. The robot-like character is called a "man-machine". The "man-machine" is Maria, an innocent and compassionate teacher of children turned temptress when her essence is captured by the wizard and incorporated into the body of the machine. In *Metropolis* (1927), the laborers live underground in the depths of the city, while the ruling elites live in luxury over them. The wizard (Rotwang) builds a female robot to demonstrate that he can use science, technology and magic to create "life". The German expressionist style of cinema often juxtaposed the images of high modernity with the mystical. Rotwang is a wizard, and his robot creation is forged in a mysterious electrified laboratory.

In *Metropolis*, laborers are represented as cogs in the machines—dispensable humans. There is one striking scene in the film where machines in the depths of the city malfunction, killing all the workers who attend to them. The dead are then thrown into the machines and new workers arrive to replace them, restarting the process all over again. Thus, human-like machines are created on the one hand, while humans become machines on the other—a dualistic paradox. Andreas Huyssen says of the 1920s period and *Metropolis*:

> The simple fact is that stylistically *Metropolis* has unusually and mainly been regarded as an expressionist film may give us a clue. And indeed, if one calls expressionism's attitude to technology . . . the film actually vacillates between two opposing views of modern technology that were both a part of the Weimar culture. The expressionist view emphasizes technology's oppressive and destructive potential and is clearly rooted in the experiences and irrepressible memories of the mechanized battle-fields of World War 1
>
> (Huyssen 1981, p. 223)

German critic Axel Eggebrecht condemned *Metropolis* as 'mystifying distortion of the unshakable dialectic of class struggle' (cited in Huyssen 1981, p. 221). This is illustrated when the film concludes on a note of reconciliation between right and left. The boundary between the laborers and machines is in crisis, made permeable in the act of production. *R.U.R.* and *Metropolis* criticize the boundary transgression and show the horror of this boundary; yet, in *Metropolis*, the machine is also romanticized and celebrated, as was common in German Expressionism (Huyssen 1981).

SEPARATE AND/OR HOMOLOGOUS RELATIONS

Karel Čapek was not a Marxist, nor was he uncritical of the practices of capitalism. Čapek rebelled against these dominant political practices and ideologies of the time. Klíma suggests this is why his plays are concerned with outsiders (2004). Haman and Trensky write that Čapek's work 'reflect[s] the problems of modern civilization and its technical explosion' (1967,

p. 176). Čapek was known for his 'moral relativism' (Haman and Trensky 1967, p. 175). My justification for using the works of Marx is motivated by the types of politics that were prevalent in the 1920s in Europe, Russia and the US. This is also supported by the themes that inspired Čapek to write *R.U.R.* and in 1908 write a short story called "The System":

> ... the story extends ad absurdum the rationalization of manufacture, turning man into a small part of the manufacturing mechanism and thereby reaching maximum productivity. But dehumanization and rationalization can only be carried so far. As soon as the workers grasp their situation, they revolt and destroy the system. That's how the 1908 story ends.
>
> (Klíma 2004, p. xii)

A character that would symbolize a human without subjectivity is indicated in "The System", and was consolidated in the figure of the robot in *R.U.R.*, 'turning man into a small part of the manufacturing mechanism' is substituted by a new character—the robot. In "The System", John Andrew Ripraton speaks on behalf of the manufacturing class he represents. This is his vision of the labor process (as articulated by Čapek):

> The labour question is holding us back. . . . The worker must become a machine, so that he can simply rotate like a wheel. . . . A worker's soul is not a machine, therefore it must be removed. This is my system. . . . I have sterilized the worker, purified him; I have destroyed in him all feelings of altruism and camaraderie, all familial poetic and transcendental feelings. . . .
>
> (cited in Klíma 2004, p. xii)

On the political right, the age of mass mechanization was characterized in the scientific production methods of Frederick Taylor (1911) and General Motors (GM) in the US. On the political left, events were inspired by the communist revolution in Russia in 1917, led by Vladimir Ilyich Ulyanov— Lenin, the leader of the Bolshevik party. I present it like a contest because both systems of thought put a huge emphasis on the mechanization of production practices.

I am interested in the points of symmetry and asymmetry between socialists and capitalists. How did socialist and capitalist practices and ideologies influence the first concept of the robot? Marx believed that advances in the production methods brought about by industrialization and advances in technology held the key to improving and transforming capitalist societies into socialist, and then communist, economies. 'Everything will be done by living machines. People will do only what they enjoy. They will live only to perfect themselves,' says Domin in *R.U.R.* (Čapek 2004, p. 21). The characters in *R.U.R.* deliberately express an outlook of the period, and Čapek almost certainly modeled Domin (the manager of Rossum's factory and utopian visionary) on the archetypal Marxist character. The practices of leisure and consumption were largely absent from the mass of ordinary people in the 1920s, while communists offered a system of organization that would allow people to develop their skills and talents. Marx and Engels wrote in one of their brief statements on communist society:

> In communist society, where nobody has one exclusive sphere of activity but each can become accomplished in any branch he wishes, society regulates the general production and thus makes it possible for me to do one thing today and another tomorrow, to hunt in the morning, fish in the afternoon, rear cattle in the evening, criticize after dinner, just as I have a mind, without ever becoming hunter, fisherman, shepherd or critic.
>
> (Marx and Engels, 1970, p. 53)

The Marxist Ilyenkov illustrates the fantasy of a communist utopia when he refers to the novel *Red Star*. Published in 1908 by Alexander Bogdanov, *Red Star* is set on Mars in a future society where work is done by machines and humans only supervise them:

> Machines do everything for the people. People only supervise them. A few hours of work where it is needed for society as a whole (indicated by the figures on the brilliant scoreboards), and you are free. What do Martians do after work? Who knows . . . Leonid N. (here they call him Lenny) isn't allowed to look into this. Perhaps they devote themselves to love, perhaps art, perhaps intellectual self-improvement. But these are everyone's private matters, and on Mars it is not acceptable to poke one's nose into private matters.
>
> (Ilyenkov 1982, p. 60)

While robotic machines will free humans for work—allowing them to be mere supervisors to machines—the humans fear that machines will replace the need for human labor and therefore increase unemployment. The widespread influence of the radical philosophies of Marxism in the 1920s gave an intellectual framework for making sense of the age of machines. In *Capital* in 1887, Marx wrote:

> The contest between the capitalist and the wage-labourer dates back to the very origin of capital. It raged on throughout the whole manufacturing period. But it is only since the introduction of machinery has the workman fought against the instrument of labour itself, the material embodiment of capital.
>
> (Marx 1974, p. 403)

The machine as 'material embodiment of capital' is reflected in movements where the instruments of labor are the target of political dissatisfaction.

The Luddites damaged machinery in the 1800s and is an early example of the conflict between worker and employer via machines. The activities of Luddites were referred to as "frame-breaking" or "machine-breaking", and the Luddites used these methods to bring pressure on their employers. As historian Thomis argues, "machine-breaking" was done by Luddites, not necessarily in hostility to the machines, but as a convenient target for attack to help their bargaining powers in disputes (Thomis 1970, p. 14). The Luddites attacked machines to promote their political agenda, not because they thought the machines were animate.

The animate/inanimate aspects of "things" is something that persisted in various forms during the 1920s inspired by the political philosophy of Marxism. In *Capital*, Marx begins his analysis of political economy with the commodity. Marx explains how the capitalist process treats the human worker in a thing-like way, as labor power is bought just like any other commodity in the capitalist production process. Reflecting on these themes and issues calls into question separations between human and nonhuman machines. In the capitalist production process, the boundaries between human and the nonhuman machine were intimately interrelated and uncertain:

> In handicrafts and manufacture, the workman makes use of a tool, in the factory; the machine makes use of him. There the movements of the instrument of labour proceed from him, here it is the movements of the machine that he must follow. In manufacture the workmen are parts of a living mechanism. In the factory we have a lifeless mechanism independent of the workman, who becomes its mere living appendage.
>
> (Marx 1974, p. 398)

In this, Marx contrasts 'handicrafts' with the 'factory system' and draws attention to the different relations that the laborer experiences in each. Under capitalism, the search for profit and surplus value means that the worker is subordinate to the system of production. The aim of a communist revolution is to socialize production, and one of the Bolsheviks' first priorities was to obliterate private property relations by collectivizing production. The Bolsheviks in Russia adopted the techniques of Taylorism, yet, as Don Van Atta argues, the implementation of these Western practices did not follow the same course (see Atta 1986). *R.U.R.* gave Čapek a means of criticizing the political practices and ideologies of the period. Yet, he was not alone in this. Marxists also saw themselves to some extent as critics of a dehumanizing system of mechanization and the mass production of use-values, which they saw as inseparable from capitalist social relations. Marx analyzed the impact of mechanization in two distinct ways. Marx wrote of machinery as it impacted the laborer and the labor process under capitalism:

> The lightening of the labour, even, becomes a sort of torture, since the machine does not free the labourer from work, but deprives the work

of all its interest. Every kind of capitalist production, in so far as it is not only a labour-process, but also a process of creating surplus-value, has this in common, that it is not the workman that employs the instruments of labour, but the instruments of labour that employ the workman. But it is only in the factory system that this inversion for the first time acquires technical and palpable reality. By means of its conversion into an automaton, the instrument of labour confronts the labour process, in the shape of capital, of dead labour, that dominates, and pumps dry, living labour-power.

(Marx 1974, pp. 398–399)

Here, Marx is interested in how the mechanization has burdened individuals by reducing the physical aspect of labor ('the labour of Sisyphus') yet also reducing its mental content. Marx goes on to write about the separation of 'intellectual' from 'manual' (1974, p. 399). In this context, Marx writes how the labor process itself appears as an inverted relation, as it 'pumps dry, living labour power'. Under capitalism, the laborer acquired a symmetrical form with the machines. Marx argued that machinery embodied human labor, and yet living labor (labor working at the machine) confronts it not as embodied human labor, but as abstract labor, subordinated to it. Marx saw the dynamic of capital as diminishing the human element in profit-focused relations. Marx's commodities were composed of two values, 'use value' and 'exchange value'. Use value, Marx argued, existed in all human societies and merely referred to the use humans attributed to the object: 'The nature of such wants, whether for instance, they spring from the stomach or from fancy makes no difference' (Marx 1974, p. 43). Exchange value, however, only existed under the capitalist mode of production. Marx saw the chaotic surplus value creating properties of capitalism as problematic, writing '. . . the relations connecting the labour of one individual with that of the rest appear, not as direct social relations between individuals at work, but as what they really are, material relations between persons and social relations between things' (1974, p. 78).

Under capitalist social relations, humans were exploited, which created the conditions that alienated humans from social life with others and created mystified human social relations. Marx characterized the capitalist economy as a system where a homologous effect occurred between person and things. This symmetry is well expressed in Marx's notion of the form of value. Marx uses the following equation to illustration the rendering of different types of things as possible of equivalence:

x commodity a = y commodity b, or x commodity a is worth y commodity b

(1974, p. 55)

In the US, Frederick Taylor proposed a new scientific approach to management techniques. Taylor's techniques undoubtedly intensified working practices for US laborers. In 1911, he wrote '. . . the greatest prosperity can exist only as the result of the greatest possible productivity of men and machines of the establishment that is, when each man and each machine are turning out the largest possible output. . . .' (Taylor 1911, p. 1). Taylor was interested in the problem of 'soldiering' and found that workers would spontaneously loaf on the job, regardless of their pay. Taylor's techniques of management aim to reduce the scope for slacking for employees (see Taylor 1911). In this, Taylor attempted to edit out this personal element, and standardize work and production to limit the autonomy of individual workers.

THE PRODUCTION LINE

General Motors was an important manufacturer in the early 1900s and spurned the term 'Fordism' associated with the industrialist Henry Ford. According to Bailes:

> American industry utilized an unprecedented level of standardization specialization, assembly line techniques, and economics of scale achieved through huge plants. The new techniques associated with Taylorism and Fordism were given particular credit by Russian economists and Soviet leaders for America's achievement of industrial superiority.
>
> (1981, p. 429)

Henry Ford was an industrialist in the automobile industry. The infamy of Ford is reflected in the novel *Brave New World* written by Aldous Huxley in 1932 (1968). Huxley uses the name of Ford to refer to the modern era: AF, or After Ford. The assembly line was a feature of this mass mechanization process and, in the assembly line, each task would be broken down into an ever-diminishing part (Bix 2000). This led many to reflect on the work as fragmented and producing mindless drudgery. In the 1920s, there was a sense that new production techniques would eliminate any skill at all on behalf of the worker—that the laborer was a mere automaton:

> The miserable routine of endless drudgery and toil in which the same mechanical process is gone through over and over again, is like the labour of Sisyphus. The burden of labour, like the rock, keeps ever failing back on the worn-out labourer.
>
> (Marx citing Engels 1974, p. 398)

Writing in the *New York Times* in the 1930s, Bertrand Russell commented on the machine. He wrote how machine technology would lead to the ever-diminishing value and independence of the individual (Ward and Lindbergh 1972, p. 75). Anthropologists Carrier and Miller write:

> Central to Marxist anthropology was the critique of what can be called a commodity culture, in which the abstraction of capital diminished the humanity of workers by reducing them to mere pawns in the strategies by which capital reproduced itself. The more that dominant thought within a society construed capital as an organic force with its own growth, agency and reproduction, the more that actual humanity was understood to lose those qualities and become merely the means for capital development.
>
> (1999, p. 33)

This led to an interesting dichotomy. On the one hand, the activity of labor was decreasing the value of the individual while, on the other, it was giving rise to (in some eyes) a homogenous mass that resulted in both fears of loss of individuality and fears of totalitarianism (Carrier and Miller 1999). The fear of individual annihilation was expressed in several forms. Gustave Le Bon expressed these fears through his metaphor of the mob. In *The Crowd,* 'the mob' was a homogenous mass where the individual was assimilated in the collective and acted like an automaton (1952). The mob that Le Bon feared was the working class, which he saw as able to be manipulated by the political left. On the left, the fear of homogenization was seen as an outcome of capitalist social relations. This was compounded by a view that the specific activity of labor was an alienating practice—which acquired further intensity with new methods of management that led to the further fragmentation of the act of production:

> The First World War, with its increased industrialization and demands for factory manufacture of standardized parts, led to greater concern that people themselves were becoming, as Ruskin feared, almost subhuman creatures devoid of emotions and individuality.
>
> (Wosk 1986, p. 149)

Soviet and capitalist systems both embraced the new technologies of production, which continued well beyond the 1920s, during the period in which Čapek was writing, as left and right saw the salvation of humanity as connected to advancing methods of manufacture. The robot is a symbolic figure that captures paradoxical feelings on the nature of the human and machine and the interrelations between them. In differing ways, artists and political activists had begun to illuminate the various ways in which the breakdown of boundaries between the human and nonhuman had occurred. This symmetry acquired particular forms in both left philosophies of Marxism and the capitalist right philosophies. Moreover, the penetrating forces of production laid bare a sense that something was lost and transgressed—individuality, skill and freedom. For Čapek, his intention was explicit and, in creating the robot, he could comment and judge the modernizing and rationalizing processes that he saw dominating life in the 1920s. As this chapter is a homage to Čapek's work on robots, I will end with his words that he wrote in 1923 in *The Saturday Review*:

> Yes, it was my passionate wish that at the moment of the Robots' attack, the audience felt that something valuable and great was at stake, namely humanity, mankind, us. . . . Yet imagine yourself standing at the grave of mankind; even the most extreme pessimist would surely realize the divine significance of this extinct species, and say it was a great thing to be human.

BIBLIOGRAPHY

Atta, DV 1986, 'Why is there no Taylorism in the Soviet Union', *Comparative Politics*, vol. 18, no. 3, pp. 327–337.

Bailes, KE 1981, 'The American connection: ideology and the transfer of American technology to the Soviet Union, 1917–1941', *Comparative Studies in Society and History*, vol. 23, no. 3, pp. 421–448.

Benson, TO 1987, 'Mysticism, materialism and the machine in Berlin dada', *Art Journal*, vol. 46, no. 1, pp. 46–55.

Bix, A 2000, *Inventing ourselves out of jobs?: America's debate over technological unemployment, 1929–1981*, Johns Hopkins University Press, Baltimore.

Bon, GL 1952, *The crowd*, Ernest Benn, London.

Čapek, J & Čapek, K 1961, *R.U.R. and the insect play*, Oxford University Press, Oxford.

Čapek, K 1924, 'Why I am not a communist', *The Pritomnost (Presence) Magazine*, December, trans. M Pokorny. Available from: <http://capek.misto.cz/english/communist.html>. [22 July 2014].

Čapek, K 1935, 'The author of robots defends himself—Karl Capek', Lidove Noviny, 9 June, trans. B Comrada.

Čapek, K 1961, *R.U.R. (Rossum's universal robots)*, trans. P Selver.

Čapek, K 1998, *War with the newts*, Penguin, London.

Čapek, K 2004, *R.U.R. (Rossum's universal robots)*, Penguin Classics, New York.

Carrier, J & Miller, D 1999, from 'Private virtue to public vice', in Anthropological theory today, ed. H Moore, Polity Press, Cambridge, pp. 24–47.

Corn, JJ & Horrigan, B 1984, *Yesterday's tomorrows: past visions of the American future*. The Johns Hopkins University Press, Baltimore.

Green, C 2006, 'The Machine' in *Modernism: designing a new world, 1914–1939*, ed. C Wilk, V & A Publications: London, pp. 71–111.

Haman, A & Trensky, PI 1967, 'Man against the absolute: the art of Karel Čapek', *The Slavic and East European Journal*, vol. 11, no. 2, pp. 168–184.

Huxley, A 1968, *Brave new world*, Heron Books, London.

Huyssen, H 1981, 'The vamp and the machine: technology and sexuality in Fritz Lang's Metropolis', *New German Critique*, no. 24/25, pp. 221–237.

Illyenkov, EV 1982, *Leninist dialectics and the metaphysics of positivism*, New Park Publications, London.

Jiménez, AC & Willerslev R 2007, 'An anthropological concept of the concept: reversibility among the Siberian Yukaghirs', *Journal of the Royal Anthropological Institute* vol. 13, no. 3, pp. 527–544.

Klíma, I 2004, 'Introduction', in *R.U.R. (Rossum's universal robots)*, Čapek, K, Penguin Classics, New York. First published in Prague 1921 by Aventinum.

Lodder C 2005, Constructive strands in Russian art, Pindar Press, London.

Marx, K 1974 [1887], *Capital*, vol. 1, Lawrence & Wishart, London.

Marx, K & Engels, F 1970, *The German ideology*, trans. , Lawrence & Wishart, London.

Reichardt, J 1978, *Robots: fact, fiction and prediction*, Thames and Hudson, London.

Smejkal, F 1990, *Devetsil: an introduction in Czech avant-garde art, architecture and design of the 1920s and 1930s*, Museum of Modern Art, Oxford & Design Museum, London.

Taylor, FT 1967, *The principles of scientific management*, W. W. Norton, New York.

Thomis, M 1970, *The Luddites: machine-breaking in regency England*, David & Charles Archon Books, Newton Abbot, England.

Vlcek, T 1990, 'Art between social crisis and utopia: the Czech contribution to the development of the avant-garde movement in east-central Europe', 1910–30, *Art Journal*, vol. 49, no. 1, pp. 28–35.

Ward, J & Lindbergh W 1972, 'Linking the future to the past' in *The social and cultural life in the 1920s*, ed. RL Davis, Holt Rinehart and Winston, New York.

Wilk, C 2006, *Modernism: designing a new world: 1914–1939*, V&A Publications, London.

Wosk, JH 1986, 'The impact of technology on the human image in art', Leonardo, vol. 19, no. 2, pp. 145–152.

Wright, WE 1966, *Serf, seigneur, and sovereign: agrarian reform in eighteenth-century Bohemia*, University of Minnesota Press, Minneapolis.

Film Cited

Metropolis 1927, dir. Fritz Lang.

2   Out of Body Minds

In the film *2001: A Space Odyssey* (1968), viewers meet a highly sophisticated artificial intelligence (AI) system known as the HAL 9000 (an acro-
nym of **H**euristically programmed **AL**gorithmic computer). The film and
this chapter explore a central issue in AI: the role of the body in artificial intelligence systems. The story is very (briefly) summarized as follows:
HAL 9000 is the on board computer of a ship accompanying astronauts
on a mission to Jupiter, where signs of intelligent life are detected. As the
journey progresses, Hal grows more peculiar, making errors and suffocating the crew in hibernation by turning off their life support systems. This
causes the two remaining crewmembers, Dr. David Bowman (Keir Dullea)
and Dr. Frank Poole (Gary Lockwood), to take back control of the ship.
They know that HAL is a threat, so they find a suitable space to engage in
a private conversation concealed from HAL, but HAL is able to read their
lips and becomes aware of their plan to disengage HAL's computer systems.
HAL 9000 is represented as a red circle enclosed within a black circle on a
computer monitor and has no ears, mouth, or eyes. Despite Hal's lack of a
body, Hal can hear and see, which leads to catastrophic results for all of the
crewmembers (including Bowman), who are left to drift off to Jupiter, alone,
with no hope of returning to Earth.

The film's themes are significant for a number of reasons. In the 1960s,
AI was a new and exciting growth area with untold potential yet to be har-
nessed. Experts predicted that, by the year 2001, AI would be solved and
machines like HAL would be present. 2001 came and went, however, and AI
systems imagined from the 1960s are still to be realized.[1] The astronauts try
to resist HAL, who is now uncontrollable, but HAL foils their plot by reading their lips and thereby short-circuiting their plan to rebel. But no one asks
how HAL managed to read their lips. Hal has no eyes; it has only a camera to
observe the world around it. But how does the machine interpret the images
that it receives? Lip-reading is a complex skill that requires knowledge of
the shape of mouths and the expulsion of air at particular points to create
sounds (Walker 1999; Nitchie 1950; Nitchie 1913). As Nitchie explained,
'The problem of teaching lip-reading is truly a psychological problem. . . .
The difficulties for the eyes to overcome are two: first, the obscurity of many

of the movements, and second, the rapidity of their formation. That spoken language is not well adapted to the purpose of lip-reading is evident from the many sounds that are formed within the mouth or even the throat. The difference between vocal and non-vocal consonants is invisible to the eye' (1913, p. 5). How could HAL read lips without a body? Which leads us to ask: what is the body for? What is the mind capable of, or incapable of, in its absence? These questions are important for us because the presence or absence of particular kinds of bodies provides the framework for disembodied AI and embodied and behavior-based robotic systems.

MILITARISM, TURING AND THINKING MACHINES

Early computing and AI grew out of Western militarism in World War II. The desire to crack codes and produce greater control of technical objects for military and strategic purposes was paramount to the conflicting European and American military operations. It was a German national, Konrad Zuse, who developed the 'first programmable computer' (Garnham 1987, p. 5). German military advances spurred the allies on with determined vigor:

> In the USA, scientists . . . at the University of Pennsylvania had developed the ENIAC, a machine for calculating bombing tables. . . . In Britain, Alan Turing and a team of cryptanalysts at Bletchley Park had used electromagnetic computing machines for code breaking.
>
> (Garnham 1987, p. 5)

A system of technology built on modern militarism has shaped the discipline and field since World War II, and even today military funding of AI, computing and robotics makes up the largest slice of research funding. Breaking computer codes and intercepting enemy military operations are driven by particular notions of the other.

The 1940s onwards saw the development of computational machines. Computation is to "compute," meaning to calculate or reckon. Charles Babbage intended to develop a computation machine, the Difference Engine, in 1822, but he never received the funding to bring the project to fruition. Babbage imagined, and partially constructed, the Analytic Engine in 1837, but once again his ideas never came to complete fruition. The first programmable computer, as stated, was developed by German Konrad Zuse (1936–1938), but the first electrical computing device was the result of a team at Bletchley Park in 1943. While that may have occurred initially in a specific location, interest in computers mushroomed and became an integral part of the war effort. Feminist writer of science and technology Donna Haraway explains how developments in radar technologies and missile systems helped to reformulate technological goals and pejoratives. It was here that the boundary between human and machine was reworked, as military victories forced nation-states to develop more sophisticated technologies:

> War and problems of military management encouraged new develop-
> ment in science. Operations research began with the Second World War
> and efforts to co-ordinate radar devices and information about enemy
> position in a total or systems ways, which conceived of the human oper-
> ator and the physical machines as the unified object of analysis.
>
> (Haraway 1991, p. 58)

The 1940s also saw new theorizing about organism-machine relationality: cybernetics. In the 1940s, Norbert Weiner developed cybernetics, an outgrowth of communications theory that fused machine and organism (Weiner
1961). Cybernetics spawned a later vision of the cyborg—part human and part
machine. As Mike Featherstone and Roger Burrows explain, cybernetics syner-
gized 'the human mind, the human body and the world of automatic machines
and attempted to reduce all three to the common denominator of control and
communications' (1995, p. 2). Cybernetics is an outgrowth of communication theory and computing, as it represented an innovative departure from
the rigid theorizing of computing modeled on scripted algorithms. Cybernetics is an organically inspired approach to computing and was influenced by a
variety of thinkers, including anthropologists, with some of its early adherents
being Margaret Mead and Gregory Bateson. The 1940s and 1950s were a
time of growth for computing, with AI soon to follow in 1956; however, while
cyberneticians were inspired by animal organisms (including humans), AI put
disembodied human cognitive processes very much to the fore.

Alan Turing's seminal paper "Can a Machine Think?" provided the
modern grounding of AI (1950). Turing was a well-respected prominent
computer scientist and World War II code-cracker based at Bletchley Park,
England, yet there was also something unsavory about Turing for the establishment. Born on 23 June 1912, he was the son of a colonial Indian civil
servant and was raised in India until his parents relocated him and his sib-
ling to England. A student at Cambridge in the 1930s, his reputation was
built from developing the computer Colossus.

Turing proposed a counter intuitive model for machine intelligence. Ignoring the appearance of the machine, Turing judged machine intelligence on the
basis of a conversation, comparing it to how a human might respond in the
same circumstances. He speculated that it would not even need to provide
a response comparable to a human's 100% of the time—70% appeared to
be an accurate enough proportion of time to convince a human interlocutor.
Arthur C. Clarke pays homage to Turing in the book *2001: A Space Odyssey*,
and we find that Turing is referred to as an intellectual ancestor of HAL 9000:

> Turing had pointed out that, if one could carry out a prolonged con-
> versation with a machine—whether by typewriter or microphone was

immaterial—without being able to distinguish between its replies and those that a human might give, then the machine was thinking, by any sensible definition of the word

(Clark 1990, p. 108)

Logical reasoning as mathematical formula was developed a century previously by George Boole. In 1854, Boole published his text *Investigation of the Laws of Thought* that described a method of thinking called "Boolean Algebra". In the book, Boole argued that logic is a form of mathematics, and, like geometry, it is based 'on a foundation of simple axioms. . . . And where arithmetic has primary functions such as addition, multiplication and division, logic can be reduced to operators such as "and", "or", "not"' (Strathern 1997, pp. 26–27). The advantage of this system is that it could work in the base-2 binary system, as opposed to the more complex decimal system, which has ten digits. Boole's work laid the foundation for imagining complex intellectual processes as mathematical.

Turing had two distinct, yet overlapping, notions about machines. The first was later coined The Turing Machine, a mathematical formula for describing the logical workings of a machine, meaning that he 'mapped out the theory of computers before a single computer (as we know it) had even been constructed' (Strathern 1997, p. 50). Turing is also renowned for his Turing Test—another set of logical propositions about what might need to be exchanged between a computer and a human being for a computer to be perceived as intelligent. Turing's formulation of what makes an intelligent machine still withstands, with thousands of dollars in prizes for any successful designer of such a program (Christian 2011). The Loebner Prize, for example, is an annual competition to test the effectiveness of such a program and, as of 2014, no one has successfully succeeded in meeting the challenge of convincing a human being they were engaging with another person. The annual event continues to be a staple of the computing calendar.

Why would this conversational aspect of human intelligence demonstrate the philosopher's stone of AI? Using the motif of disguised physical appearance, it is the capacity to convince the human counterpart that they are engaging with an intelligent entity. Alan Turing, notably, set in place a model for which to compare and contrast humans with machines. Turing primarily took the notion of a "conversation" and explored why he might have believed this to be foundational in the development of machine intelligence.

What is a conversation? A conversation is a type of interpersonal exchange involving the participants' communication via natural language. A conversation is arguably a unit of interpersonal exchange that acts as a marker of recognition between different agents (usually persons, but not exclusively so) who are able to use a common language to communicate their thoughts, feelings and intentions. Conversational exchange is mediated by one or more persons through speech, the body and text, and a conversation is something that is transitory, stopping and starting as it approaches each specific speech and language context. Sociologist of technology Sherry Turkle believes that spoken conversation is under threat in the digital computing age, as users prefer to relate to each other in absence rather than face-to-face (Turkle 2011). Language is neither 'affect-free, context-independent propositions', but instead, speech is a 'situated social activity' with 'the conventional dichotomy between language and non-verbal communication . . . less secure' (Ingold 1996, p. 6). Conversations can take place between persons even when these persons are not physically present (as in a telephone conversation), where the auditory and spoken aspects of communication are detached from the physically present embodied form. Conversations are ephemeral in the sense they start and stop, and there is an element of unpredictability and innovation in a conversational exchange (Chomsky 1972). Human and animal bodies are also arguably a complex language system (Birdwhistell 1971; Darwin 1998).

Turing introduced the concept of a thinking machine in relation to a 'sexual guessing game'. Turing begins his paper 'Computing Machinery and Intelligence', published in 1950 in the journal *Mind*, with a discussion of the Imitation Game. The Imitation Game is played with three parties, a man (A), a woman (B), and an interrogator (C) who can be of either sex. C only knows A and B in terms of X or Y and must say at the end of the game that A is X or B is X. Imagine, in this scenario, that the interrogator is human and A is a machine. The interrogator stays in a room separate from A and B and, by way of feeds on a paper, has to determine the sex of A and B depending on the answers C (the interrogator) receives. The aim of the game is to determine if A or B is male or female, but A and B have some interest in telling the truth in a concealed way to confused the interrogator. For example, if C (the interrogator) asks A (the male) how long is your hair? A's answer will be to confuse C to make the wrong identification, so A's answer might be 'My hair is shingled, and the longest strands are about nine inches long' (1950, p. 434). So that C cannot get clues from handwritten communications in the game, 'The ideal arrangement is to have a teleprinter communicating between the two rooms,' wrote Turing (1950, p. 434). Turing then proposes that A be replaced by a machine:

'What will happen when a machine takes the part of A in this game?' Will the interrogator decide wrongly as often when the game is played like this as he does when the game is played between a man and a woman? These questions replace our original, 'Can machines think?'

(Turing, 1950, p. 434)

In reposing the question in terms of machines, he inadvertently diminishes the importance of the body, or particularly a robot body:

The new problem has the advantage of drawing a fairly sharp line between the physical and the intellectual capacities of a man. No engineer or chemist claims to be able to produce a material that is indistinguishable from the human skin. It is possible that at some time this might be done, but even supposing this invention available we should feel there was little point in trying to make a 'thinking machine' more human by dressing it up in such artificial flesh.

(Turing 1950, p. 434)

This passage is telling, as it reveals Turing's belief that there was 'little point in trying to make a "thinking machine" more human by dressing it up . . . in artificial flesh'. This remark informs us that Turing did not see the mind as an embodied mind, but as a disembodied cognitive system. For Turing, then, he separated appearance and thinking capabilities. He also emphasized the role of deception and conversation, appearance rather than essence. Turing's framing of the problem of machine intelligence in terms of deception and lacking a body is revealing. This divide between the importance and non-importance of a body continues into the present. Turing's interest in deception is undoubtedly connected to his presence on a project deciphering secret codes, but also living a secret, alternative life due to his sexuality. Strathern (1997) explains thus '. . . this identification with a machine soon began to permeate his whole life. To regard himself as a machine provided a great psychological relief from the continuing turmoil of his inner life' (p. 73). Turing later told a future partner, Neville Johnson, 'I have more contact with this bed, than with other people' (Strathern 1997, p. 74).

Despite Turing's exceptional capabilities as a member of the elite group at Bletchley Park, his sexuality was something of a problem and embarrassment for the British establishment. In 1952, he was charged with gross indecency under Section 11 of the Criminal Law Amendment Act 1885. Two years later, in 1956, Turing took his own life with cyanide, a deadly compound that brings about agonizing pain in the receiver and a relatively quick (the recipient may continue breathing for moments after digestion) but painful death. In a letter he wrote to his friend Norman Routledge in the 1950s, he confirms that he had been in trouble with the authorities over his attraction to men. Writing at the conclusion of the letter:

> Turing believes machines think
> Turing lies with men
> Therefore machines do not think

He ended the letter, 'Yours in distress, Alan'.

(p. xiii)

Turing's otherness, his alterity and difference, was influential in his theorizing about machine thinking—his life must have seemed like an imitation game of sorts, with revelations and secrets, coding and decoding. Turing also separated thinking from a body, whilst A and B's bodies were gendered, their thinking about their bodies was detached and incompatible, even deceptively substituting one form for another.

BODIES AND MACHINES

Artificial intelligence as a field emerged as the result of a research workshop in Dartmouth in 1956. It was here that the early pioneers of AI, Marvin Minsky and John McCarthy, attended college, and where the term 'artificial intelligence' was coined (by McCarthy). What initially began as a research conference to signal a new field quickly transpired into a research project, with the first lab established at MIT. MIT is an important site for the development of early AI.

As a result of this six-week conference, McCarthy and Minsky jointly established the former Artificial Intelligence Laboratory at MIT, now the Computer Science and Artificial Intelligence Department (CSAIL). Initially comprising only a few researchers, the field quickly grew, with more than a dozen outlets emerging by the late 1960s. The recently deceased John McCarthy was based at Stanford Artificial Intelligence Laboratory (SAIL)— the lab that he founded in the late 1960s.

In the heyday of AI, from the 1960s to the early 1970s, some of its leading thinkers broadcasted their optimism about the research area. In 1965, Herbert Simon (one of the original attendees of the Dartmouth conference) wrote 'Machines will be capable, within twenty years, of doing any work a man can do'. Minsky infamously proclaimed, in 1967, that 'Within a generation, I am convinced, few compartments of intellect will remain outside the machine's realm, the problems of creating "artificial intelligence" will be substantially solved' (Minsky 1967, p. 2). This celebratory optimism about the field continues to the present day and is expressed by a new generation of thinking in the field of AI and robotics (Moravec 1998; Kurzweil 2013; Kurzweil 2000). This imagination of the impossible is interlaced with technological scenarios of the future.

The 1980s saw a lull in the breakthrough of traditional AI developments and, rather than being on course to build a HAL 9000, it was a field that was floundering. Funding was being withdrawn and problems that up to then had been the focus of AI researchers were being replaced with more practical issues, such as complex automated systems rather than reproducing an AI copy of "man". Its early pioneers' claims of AI as a radically transformative field of technology were undermined by the field's own failures. So optimistic was this field that the following story is recounted to express this optimism and its failure:

In the early 1960s, when the AI was first established, its early pioneers gave the problem of machine vision to an undergraduate as part of the undergraduate research opportunity (UROP). The project was expected to only take a summer, and still today many problems of machine vision are not solved.

(Personal communication 2004)

For our purposes now, we are interested in the way in which the presence or absence of the body was framed as a central feature in developing AI systems. The following explanation was given to lay persons and visitors to explain the difference in approaches between embodiment and autonomous AI and classical and traditional AI. For the sake of brevity, from now on I will refer to embodied AI and traditional AI when describing these oppositional approaches:

> Conventional approaches to AI—such as those of the traditional model—fail as they attempt to map all the elements of the world and put them into a machine. This approach to AI fails as if there is some anomaly that occurs that has not been programmed, the machine is unable to respond to the change. Take for example a robot. In conventional mobile robotics, the environment would be mapped out and fed into the machine, but if one element presented itself that was not there before, then the machine could not adapt. Behavior-based and embodied robotics focuses instead on getting the machine to carry out its functions using its sensors and behaviours, independent of the environment.
>
> (Personal communication 2004)

The explanation above is featured cogently in scientific terms by embodied robotic scientist Rodney Brooks:

> In traditional AI there are many classes of research contributions (as distinct from application deployment). Two of the most popular are described here. One is to provide a formalism that is consistent for some level of description of some aspect of the world, for example, qualitative physics, stereotyped interactions between speakers, or categorizations or taxonomies of animals. . . . A second class of research takes some input representation of some aspects of a situation in the world, and makes a prediction. For example, it might be in the form of a plan to effect some change in the world, in the form of a drawing of an analogy with some schema in a library in order to deduce some non-obvious fact, or it might be in the form of providing some expert-level advice. These research contributions do not have to be tested in situated systems—there is an implicit understanding amongst researcher about what is reasonable to 'tell' the systems in the input data.
>
> (Brooks 1999, p. 74)

Embodiment is a system of cognition where spatiality, emotion and social-relationality play an important part (Lakoff and Johnson 1980, p. 1999;

Gibbs 2006). Embodiment in the field of 'cognitive science refers to an understanding of an agent's own body in its everyday, situated cognition' (Gibbs 2006, p. 1). The artificial construction of a mind-body dichotomy has shaped conceptual theorizing since Descartes, imposing limitations on conceiving relations between mind and body outside these forms. Cognitive psychologists, linguists and roboticists established the importance of 'kinesthetic action in theoretical accounts of how people perceive, learn, think, experience emotions and consciousness and use language' (Gibbs 2006, p. 3). Put another way, traditional AI systems rely on representing, modeling and mapping the scope of an intelligent system, designing models and modifying the systems based on algorithms. This approach is successful in capturing the rules of what Dreyfus (1992) described as 'context-independent' systems—those systems that do not rely on the phenomenal aspects of the material and social environment. Mathematical systems and traffic control systems are two examples of where traditional AI is successful. Once unpredictability is introduced into the system the model cannot adapt. In light of these types of challenges, traditional AI systems face multiple meaning disorders; they must follow a set path or a predefined course to operate (Dreyfus and Dreyfus 1986).

Within a few days of carrying out my fieldwork ethnography at MIT's AI robot lab, an intellectual difference in approaches to AI was brought to the fore. The difference is thus: traditional AI regards the physical body as unimportant or secondary to the development of intelligent systems. Intelligence is an outcome of formalistic rationality that can be translated into mathematic formulas and programs (machine language). Traditional AI is contrasted with behavior-based embodied AI with its bodily-focused emphasis on intelligence. For a machine to be intelligent, it must draw on its body in that intelligence. Intelligent action is possible for the machine. In the 1980s, robotic scientist Rodney Brooks's ideas challenged the field of traditional AI in the direction of behavior-based embodied robotics (Brooks 1999). The difference in perspective can be associated with two prominent figures in AI, both holding offices at MIT.

BEHAVIOR-BASED ROBOTICS

In the 1980s, a new generation of AI scientists emerged and robots and bodies entered the fold in a new way. The debates about reason versus body are also echoed in the debates that correspondingly took place in the mid-1980s in the social sciences and humanities (Csordas 1999; Lakoff and Johnson 1999), as the disciplinary practices of the social sciences and humanities also began to question the notion of the reason-based human being. As Csordas wrote:

It is probably no fluke of intellectual history that a turn toward the body in contemporary scholarship in the human sciences has coincided with the realization that the postmodern condition is now the uneasy condition of all intellectual activity.

(1994, p. xi)

Csordas continues with an explanation of why this might be: 'If behind the turn to the body lay the implicit hope that it would be the stable center in a world of decentered meanings, it has only led to the discovery that the essential characteristic of embodiment is existential indeterminacy' (1994, p. xi). Csordas proposes, then, that the 'turn to the body' was a response to the aftermath of disorientation resulting from postmodern philosophies. Yet, as Csordas notes, rather than providing a stable set of meanings, the body was also reduced and dismantled. If the 1980s saw an emphasis on the body, it paralleled a critical assault on reason in a number of academic fields.

Lakoff and Johnson outlined how the mind was embodied, largely built out of an unconscious consciousness and intimately attached to the human brain and body, with the sensorimotor system being a key feature of this. 'The phenomenological person, who through phenomenological introspection alone can discover everything there is to know about the mind as the nature of experience, is a fiction' (1999, p. 5) The aptly titled *Philosophy in the Flesh: The Embodied Mind and Its Challenge to Western Thought* (1999) by Lakoff and Johnson presents more than an argument for the body—it is additionally an attack on Western rationality, in that 'Our understanding of what the mind is matters deeply. Our most basic philosophical beliefs are tied inextricably to our view of reason. Reason has been taken for over two millennia as the defining characteristic of human beings. . . . It is surprising to discover, on the basis of empirical research, that human rationality is not at all what the Western philosophical tradition has held it to be' (1999, p. 3–4).

Embodied intelligence needs a body, a physical container within which to interact with the environment. But embodied AI is more than just an argument for a physical body—computers, after all, have bodies of sorts, even if immobile and nonhumanoid. In embodied AI, the emphasis on the body extended beyond the mere recognition that an AI entity must have a body, but the recognition that intelligence, cognition, perception, linguistic and nonlinguistic communication is only possible through bodies—sensual, fleshy, sensory-motor, proprioceptive bodies. Experiments in perception and sociable robotics in this robotics lab were two main areas that emphasized this theme of bodily-based intelligence.

To complement the embodied AI arguments I heard regularly in the lab, I decided to attend a series of seminars held by Minsky at the MIT Media Lab. The Media Lab is an important technological center at MIT, and its researchers have a reputation for re-imagining human interactions with the built environment and new technologies. The large auditorium was the setting for a series of lectures in 2003 by Minsky designed to explore ideas in his then-forthcoming book *The Emotion Machine* that was later published in 2007. As one of many attendees scattered around the fixed chairs, I listened with interest to his ideas from the book, though the topic of emotion was presented to me in an unfamiliar manner. I was taken by his strength of character and his display of warmth and interest in the young undergraduates that asked him questions. My question, however, was not answered with such warmth and enthusiasm. I asked Minsky about the role of the body in developing emotional systems. Minksy replied, almost in a stern manner, to my question:

> If the body was paramount, how were disabled people or those with limited control over their bodies able to think?

Minsky raises an important theme about the body's importance in intelligent systems. The human body is varied, with multiple ways of being and senses distributed differently through each human body (Mol 2002). In disabled bodies where movement was limited, humans were still able to think. At the same time, disabled bodies are still bodied; disabled bodies are different, but no human being exists as a brain on a stem—each human body is a unique configuration (Ginsburg and Rapp 2013). The analogies between disabled bodies and machine were a recurring theme. Minsky was drawing on the Cartesian model of mind separate from body to make a case for disembodied intelligence, and inadvertently for disabled persons who may not be able to fully use their bodies. This reveals an error, because, in whatever form the disabled body is, it is still a body: every human has a body, a complex body made up of many parts (Mol 2002). Embodied intelligence is the spatial-sensual organization of experience at the level of being. Metaphor is a useful way to explore this issue. Linguists Lakoff and Johnson (1980) examined the role of metaphor in the formation of intelligence; for example, war metaphors used in speech, such as "He got shot down". For Lakoff, the development of mind was related to metaphor. *Women, Fire and Dangerous Things* (1987) is a study of the role of metaphor and embodiment in the development of categories. The title of the book refers to the category of "balan", which includes women, fire and dangerous things—which, in linguistic terms, cuts across several categorical boundaries. Balan is a term from the Australian aboriginal language Dyirbal (Lakoff 1987, p. 5). The authors argue that concepts are intimately tied to physical, social and cultural architectures, and are not an outcome of pure metaphysical concepts that can be determined and classified in logical and disembodied terms.

THE OTHER: ANIMALS AND MACHINES

The desire to produce human-like qualities in AI machines and robots may shake certain assumptions about what it means to be human. While animals were notably used as the comparative marker of distinctive humanity, machines (computer systems and robots) arguably have also become the other through which to reflect on the human. Attitudes on the boundaries between humans and animals provide a useful barometer by which to explore potentially changing attitudes towards humans, AI machines and robots. Throughout the last two centuries, it was animals that often provided the "mirror" for what was distinctive about humans. The desire to define human uniqueness is discussed through what was traditionally thought its opposite—non-human animals. Human capabilities, traits, and values were measured in relation to the animal world. Primates, for Haraway, allowed her to reflect on the socially-constructed nature of being human via creatures of the "boundary creatures" when she writes, 'Monkeys and apes have a privileged relation to nature and culture for western people; simians occupy the border zones between those potent mythic poles' (1992, p. 1). Writing in the 1970s, Haraway examines how "humanness" is articulated through animal otherness:

> People like to look at animals, even to learn from them about human beings and human society. People in the twentieth century have been no exception. We find the themes of modern America reflected in detail in the bodies and lives of animals. We polish an animal mirror to look for ourselves.
>
> (1978, p. 37)

Haraway's project examines how gender and race are constructed through the study of primates, analyzing the reflections via the processes of 'sex and economics, reproduction and production' (1978, p. 37). Today, Haraway's work has taken a more radical turn with a call for relational ontologies. In this sense, animals do not act as a "mirror" of humanity: humans and dogs radically remake each other (2003).

Animals were exemplars of diminished otherness and humans (specific, cultural-inspired classes) were "superior" to animals because of language, consciousness, agency, culture and art (Malik 2000; Kranzt 2002). Campaigners such as philosopher Peter Singer, who argued that animals deserve the same respect and status as that attributed to human beings, now want to speculate about this in relation to robots (Lin, Abney & Bekey 2012). "Culture" was also seen as the distinguishing feature of humans from non-humans. According to Wolfe, Max Weber 'viewed culture as man's emancipation from the organically prescribed cycle of nature life' (cited in Wolfe 1991, p. 1073), while, for Durkheim, 'it is civilisation that has made man what he is; it is what distinguishes him from the animal: man is man only because he is civilised' (cited in Wolfe 1991, p. 1073). In this context, 'civilisation' was viewed as artificial and human-made, and this was one of the features that distinguished humans from animals. Weber and Durkheim are not only citing that animals are the other, but are making a value judgment on how that distinction elevates humans and human society. Animals and machines have never been quite so separate.

AI continues the mechanistic philosophies of the human and so continues a lineage of practices and philosophies of the body of the seventeenth and eighteenth centuries in European thought. In the seventeenth century, the scientist Marcello Malpighi wrote 'the mechanisms of our bodies are composed of string, thread, beams, levers, cloth, flowing fluids, cisterns, ducts, filters, sieves, and other similar mechanisms' (Malik 2000, p. 36). In 1794, in a book entitled *L'Homme Machine*, the French doctor Julien Offray de la Mettrie believed a human being to be machine-like. 'The title, "machine man", refers specifically to the Cartesian hypothesis that animals are mere machines without a soul, and La Mettrie claims that what he is doing is simply applying the Cartesian hypothesis to humans'; he shows repeatedly that whatever applies to animals applies equally to humans' (Thompson 1996, p. xvi).

The practice of AI stimulated a new approach to reviewing the boundaries between human, animal and machines:

> Just as Darwinian theory stimulated sociology by posing the question of whether there were capacities that—in contrast to animal behavior— were uniquely and specifically human, recent research in artificial intelligence (AI) cognitive science, and neurobiology raises exactly the same question—only this time in contrast to machine behavior. Artificial intelligence can be viewed as a Gedanken experiment, an effort to pose a series of interrelated 'what if' questions. The fascination with AI is surely due to the philosophical issue raised by the possibility that machines can carry out activities once thought to be exclusively human.
> (Wolfe 1991, p. 1073)

AI practices have done as Wolfe suggests, in that they provide a means to examine and redraw the boundaries between human and machine. *In Man, Beast and Zombie: What Science Can and Cannot Tell Us about What It Means to Be Human*, Kenan Malik assesses the importance of the Western scientific tradition on altering conceptions of what it means to be human, particularly the influence of Descartes. Scientific-rationalism underscores the modern project and modernity-altered concepts of what it meant to be human. For example, Farnell (1994) has challenged these sharp dichotomies and discusses what she calls the:

> reconstruction of classical precepts about the nature and role of person and agency and the dualistic thinking that has not only separated body from mind, but also created opposition between subjective and objective, mental and material-behavioural, thinking and feeling, rational and emotional, and verbal and nonverbal.
> (p. 930)

Helmreich locates AI theorizing as an outcome of an outcome of Cartesian dualism:

> the grand project of Artificial Intelligence was to create artifactual minds equal to or better than human minds. This project was the apotheosis of

Cartesianist thinking. If the project of Artificial Intelligence were successful, the body could be permanently split from the mind. The mind would finally be liberated from the sticky, limited, and overly emotional flesh. One could easily argue that Artificial Intelligence was conceived in the image of the rational white European male, ideally a calculating, objective, and reasonably entity, constructed in large part by naming others as subjective and irrational.

(1998, p. 131)

Yet, does the "grand project" of AI also call into question the Cartesian "indivisibility" of the human mind by assuming it can be "divisible" and its parts broken up and simulated in a machine? In AI, the mind is broken up into parts and simulated; thus, when machines mimic the capacities of the human mind, does this make the machine analogous or homologous? To charge AI theorists with a Cartesian brush is to misunderstand Cartesian models of the mind and body. Descartes asserted that the mind was unique, incapable of extension, and different from the body. Therefore the extension of the mind into machine is somewhat debatable.

Descartes views the body as a machine and not the mind:

> if I consider the body of a man as being a sort of machine so built up and composed of nerves, muscles, veins, blood and skin, that though there were no mind in it at all, it would not cease to have the same motions as at present, exception being made of those movements which are due to the direction of the will, and in consequence depend upon the mind (as opposed to those which operate by the disposition of the organs). . . .

(1993, p. 96)

Cartesian thinking is dualistic, but AI theorizing is not exclusively dualistic in the Cartesian sense. In Cartesian philosophy, the body and the mind are separate. The body is for Descartes 'divisible', acting as a container of the mind and carrying no intrinsic importance (except for imparting sense impressions to the mind). In *Meditations,* he describes the feelings of pain in parts of the body lost through amputations (Descartes 1993, p. 90).

Descartes saw the mind as the source of reason, and 'indivisible' (1993, p. 97), hence the focus of human uniqueness and difference. For Descartes, the human body possessed attributes that were similar to animals and machines. The body, unlike the mind, was mechanical in its nature. Writing in the seventeenth century, Descartes wrote in *Meditations*:

> I possess a body with which I am very intimately conjoined, yet because, on the one side, I have a clear and distinct idea of myself inasmuch as I am only a thinking and unextended thing, and as, on the other, I possess a distinct idea of a body, inasmuch as it is only an extended and

unthinking thing, it is certain that this I (that is to say, my soul by which I am what I am), is entirely and absolutely distinct from my body, and can exist without it.

(Descartes 1993, p. 91)

In a later passage of *Meditations* he explains his position further:

. . . there is a great difference between mind and body, inasmuch as body is by nature always divisible, and the mind is entirely indivisible. For, as a matter of fact, when I consider the mind, that is to say, myself inasmuch as I am only a thinking thing, I cannot distinguish in myself any parts, but apprehend myself to be clearly one and entire; and although the whole mind seems to be united to the whole body, yet if a foot, or an arm, or some other part, is separated from my body, I am aware that nothing has been taken from my mind.

(Descartes 1993, p. 97)

Descartes diminished the importance of the body rather than ignoring it, writing 'But there is nothing which this nature teaches me more expressly (nor more sensibly) than that I have a body which is adversely affected when I feel pain, which has need for food or drink when I experience the feelings of hunger and thirst, and so on; nor can I doubt there being some truth in all this' (Descartes 1993, p. 93). Descartes' arguments were a response to the Aristotelian view of the interconnectedness of nature and being. During the course of the sixteenth and seventeenth centuries, the Aristotelian framework of a purposeful universe gave way to 'purposeless particles' (Malik 2000, p. 33).

Traditional AI theorists separate the mind from the body, but using another more mechanistic approach to the mind (and body) than Descartes suggested. Linguists Lakoff and Johnson sum up the position of Descartes more in keeping with his philosophies. Lakoff and Johnson explain the importance of Cartesian dualism in influencing Western conceptions of the mind and body in that 'there is no Cartesian dualistic person, with a mind separate from and independent of the body,' arguing instead, '. . . Rather, the mind is inherently embodied' (1994, p. 5). Malik takes up this theme, writing that 'Ryle's behaviorism leads to the same problem as Descartes' dualism. Descartes' separation of mind and matter meant that it was difficult to know how mind acted upon matter, and hence how beliefs or desires led to behavior. Ryle's collapse of mind into matter, of beliefs into behaviors, makes it similarly difficult to know how one acts upon, or causes, the other' (2000, p. 314). In this sense, it is one extreme or another, with Descartes' separation starkly contrasted with Ryle's conflation.

Malik, too, cites the influence of behaviorism as posing an earlier challenge to the Cartesian divide (Malik 2000). Malik draws on Gilbert Ryle's *Concept of Mind*, written in 1949. Ryle's work was a challenge to the 'official doctrine' of mentalism that had begun with Descartes, 'the dogma of the Ghost in the Machine' (p. 313). Malik summarizes Ryle's position as thus:

> According to Ryle, talk of mind involves what he called a 'category mistake'. There is no problem in talking about 'minds' as a generality, he argued, but there is a problem in assuming that the mind exists in a specific location in our heads. It is the same kind of mistake that tourists often make in Oxford, when, after having been taken on a tour of university buildings, lawns and people, they demand to know where the university is. To talk of the mind as if it had a separate existence, Ryle argued, was mistakenly to treat an abstract characterization of a set of brain processes as if it were itself one of those processes.
>
> (2000, pp. 312–313)

In abandoning a reason-based model, the mind is no longer singled out as uniquely human or transcendent amongst embodied philosophers and robotic scientists. The following is an example of how the body is inextricably linked to AI for embodied robotic scientists. I asked the professor of the robotics lab to tell me about the making of human-like robots. The answer I was given was intimately connected to the body:

> Is it realistic? I think, that in principle, a human-like thing, entity, could be built out of different sorts of material. I may be wrong but my current understanding of the universe says in principle it can be done. And in principle I see no reason why it couldn't be done out of silicon and steel. Then there's a question of whether us humans are smart enough to do that. So, if you look at a raccoon it's very dexterous with it hands. They can untie knots and play with locks and doors and stuff. But we wouldn't think they're smart enough to build a mechanical raccoon, even though they might have the manual dexterity. Though, in principle, a raccoon could build a vacuum cleaner (laughs). But they're not smart enough. But a vacuum cleaner can be built, they have the manual dexterity to do it. In terms of the ability. . . . You know, because you can see from it untying knots and doing stuff, but there's no way it's going to because it doesn't have the mental capacity to do it. The hands are good enough. Now I believe it's possible to build a human-like entity, which is not made out of flesh. It's possible it could exist made out of silicon and steel. So the analogy is that a vacuum cleaner could exist, but a raccoon is not smart enough to actually build one. It is two separate questions. It could be made out of steel, but are we smart enough to build one?
>
> (interview with Professor Patrick Kane,
> robotics professor, pseudonym provided)

Professor Kane sees the success of AI as linked to the body of its creators—or, put another way, a smart species has to know how to reinvent itself and have the physical capabilities to do it. In the passage, Professor Kane's explanation that humans have particular kinds of bodies that are different from raccoons is underscored in the meaning he then gives to what they can do. The body may constrain the mind, but are the products of human activity merely a result of their opposable digits? In this sense, the successes as well as the limits of raccoons are linked to their bodies. But is there really a cause-effect relation between the body and the products of it? Professor Kane illustrates the mechanist behavior-based viewpoint further:

> We're smart enough to build vacuum cleaners. But there could be a fundamental theorem of the universe like Gödel's incompleteness theorem that no creature is smart enough to build a copy of itself. There could be some universal law, like that, I suspect not, but I don't know that. Since we were built out of random occurrences then I suspect that we can figure out how to harness those complex processes, then I think we are capable. But there could be those guys up in Alpha Centuri saying what those cute little humans are; oh they think they're going to build a humanoid robot. Ha ha ha! They'll never do that and they could be right . . . they may be smart enough to build a human. . . . You see I did think that we are incredibly arrogant. We are, we think we are smart enough to understand everything. And we may not be smart enough to understand everything . . . but maybe not. We are trapped like raccoons are trapped, never to be as good as humans. Maybe we're trapped, never to be good as those imaginary beings in Alpha Centuri . . . who happen to have evolved differently to us but are much smarter.

Yet Professor Kane also talks of 'being smart' as a factor in the processes of intelligence. Therefore, he has not exclusively abandoned the importance of the mind—but tied it mechanically to the body.

What are the successes and drawbacks of reason-based AI? According to Hans Moravec, 'Artificial Intelligence has successfully imitated the conscious surface of rational thought, and in doing so made evident the vast unplumbed sea of unconscious processes below' (1998, p. 24). Is it only possible to reproduce what is known and can be modeled? The traditional model has been criticized for its 'top-down' approach to intelligence. Moravec does not abandon the top-down-reason-based approach, arguing instead 'Building truly intelligent machines means exploring the ocean, from its rational top to its adaptive bottom' (1998, p. 24). Brooks, too, agrees that 'computers are now better at doing symbolic algebra . . . better at designing classes of numerical computer programs. . . . In short, they are better at many things that were previously the purview of highly trained mathematically inclined human experts' (2002, p. 170). Philosophers Hubert Dreyfus (1992) and Harry Collins (1998) have outlined the challenges that AI would have to meet in order to reach the goal of AI. Collins believes, like many in the field, that it is possible to extend the range of machine-like intelligence if we can find appropriate models of human behavior, writing, 'we can extend our robot-like activity without limits' (Dreyfus 1992, p. 717). But what are these appropriate models? Dreyfus is more interested in the activities that have met with the most success in AI. Yet, there are problems, he explains, as whether or not human behavior can be replicated in a machine is

> dependent on whether the human like behavior can be 'digitalized' that is, domains [that] have a structure that can be captured in this theoretical way as opposed to domains [that] do not have a structure that can be captured in a context-free [way] related by laws or rules.
>
> (1992, p. 717)

Dreyfus's theory of 'digitalization' is a useful way of thinking about the debate between human uniqueness versus machine simulation of human intelligence. For a human behavior to become digitalized requires that the behavior be captured with the help of theoretical rules and structure independent of context. Dreyfus is a critic of AI's ability to meet all its problems:

> Plane loading can be captured by a combination of geometry and trial and error, spectrography has a theoretical structure, and computer component choice can be carried out in a context-free way by using look-up tables and heuristic rules for how to combine components having various capabilities. Thus not all expertise is intuitive, but that is not because some expert behavior can be converted in to behavior-specific action and other expert behavior cannot; rather there is a theory of some expert domains, and when a theory is available the expert (and the computer) can use the theory rather than getting along using only pattern recognition. Experts in these fields were replaceable because they were behaving in a digitalizable way, but they were behaving digitally because the domain had a digitalizable structure.
>
> (1992, p. 721)

While Dreyfus is interested in aspects of human behavior that can be 'digitalized', his argument suggests that some human behavior is susceptible to mechanization or mental modeling. Yet Dreyfus also makes a distinction between different types of expert behavior, not all of which can be formulated in rules. In this sense, Lakoff and Johnson's argument that 'thought is mostly unconscious' and 'abstract concepts are largely metaphorical' (1999, p. 1) means that there are limits to reproducing AI based on reason-based rule formulation and perhaps even reproducing AI at all, as 'the mind is [also] inherently embodied' (1999, p. 1). In AI, the defeat of chess champion Gerry Kasparov by IBM's Deep Blue chess-playing computer in 1997 was seen as both a challenge to and success of traditional AI models. Chess is a rule-based game, yet the champions master the rules and rely on intuition and experience—which cannot be easily classified. If it is intuition that accounts for a chess champion's success, then a computer that can beat a human is able to go beyond its rule-based programs, or that intuition is itself rule-based.

In this chapter I have tried to focus the reader's attention on the types of practices that go into the making of AI embodied and disembodied systems, and the conflicts between them. Behavior-based robotics does not see the body as a mere container for a logically constructed mind, but intimately connected to the body that it encompasses, and this body contributes to the intelligence of the machine. I also tried to show how intimately militarism is bound up with the development of AI. Much like our robots in chapter one that revolted against their human masters, it is not so science fiction to think that machines are destroyers, as this is precisely what they are and are created to become. In drawing on the biography of Turing, we see how his own detachment was bound up with his creative imagination in early computing.

NOTE

1 In 2011, the IBM supercomputer Watson made its public debut on an American television quiz show called *Jeopardy!*. IBM, the makers of Watson, wanted to show off the potential of their new supercomputer by introducing it as a contestant on the game show. The *Jeopardy!* quiz format works by asking questions in response to answers. Panelists compete for cash and other prizes by guessing the right question. IBM's designers were ready to prove Watson's intelligent capacities. The difference between this and other supercomputers was Watson's ability to understand natural language. The spoken answers were reformulated as text-based ones and fed into its large database of knowledge. Language, particularly spoken language, is often opposed to scripted, formulaic language because it is hard to reproduce in machines as machines operate within precise parameters of interpretation, comprehension and expression. The crucial aspect of Watson's success was that it was able to discern the 'right' question on enough occasions to successfully win the competition. As its official website explains:

> The challenge in building a computer system like Watson lies in developing its ability to understand the language of a clue, register the intent of a question, scour millions of lines of human language, and return a single, precise answer – in less than three seconds.
>
> (Official Watson IBM Website)

Watson, developed by IBM's ProjectQA, was hailed as a success in this area, though with modifications, and showed the potential of AI machines. Watson was a by-product of both the theorizing in computing and AI. Is the debut of Watson and indication of the realization of the AI dream, first fictionalized in *2001: A Space Odyssey?*

BIBLIOGRAPHY

Birdwhistell, RL 1971, *Kinesics and context: essays on body-motion communication*, Allen Lane, The Penguin Press, London.

Brooks, R 1999, *Cambrian intelligence: the early history of the new AI*, MIT Press, Cambridge, Mass.

Brooks, R 2002, *Flesh and machines: how robots will change us*, Pantheon Books, New York.

Brooks, R & Stein, LA 1993, 'Building brains for bodies', *MIT AI lab memo* #1439.

Chomsky N 1972, *Studies on semantics in generative grammar*, Mouton, The Hague.

Christian B 2011, *The most human human: a defence of humanity in the age of the computer*, Viking, London.

Clark, AC 1990, *2001: a space odyssey*, Legend, London.

Collins, HM 1998, 'Socialness and the undersocialized conception of society', *Science, Technology & Human Values*, vol. 23, no. 4.

Csordas, T 1994 'Introduction' in *Embodiment and experience: the existential ground of culture and self*, ed. T Csordas, Cambridge University Press, Cambridge, pp. 494–516.

Csordas, T 1999, 'The body's career in anthropology' in *Anthropological theory today*, ed. H Moore, Polity Press, London, pp. 172–205.

Darwin, C 1998, *The expression of the emotions in man and animals*, 3rd edn. HarperCollins, London.

Descartes, R 1993, 'Meditations on first philosophy' in *Focus*, ed. S Tweyman, Routledge, New York.

Dreyfus, HL 1992, 'Response to Collins, artificial experts', *Social Studies of Science*, vol. 22, no. 4, pp. 717–726.

Dreyfus, HL & Dreyfus, SE 1986, *Mind over machine: the power of human intuition and expertise in the era of the computer*, The Free Press, New York.

Farnell, B 1994 'Ethno-graphics and the moving body', *MAN, Journal of the Royal Anthropological Institute*, vol. 29, no. 4, pp. 929–974.

Fukuyama, F 2002, *Our posthuman future: consequences of the biotechnology revolution*, Profile Books Ltd, London.

Garnham, A 1987, *Artificial intelligence: an introduction*, Routledge & Kegan Paul, London.

Gibbs, RW 2006, *Embodiment and cognitive science*, Cambridge University Press, Cambridge.

Ginsburg, F & Rapp, R 2013, 'Disability worlds', *Annual Review of Anthropology*, vol. 42, pp. 53–68.

Haraway, DJ 1991, *Simians, cyborgs, and women: the reinvention of nature*, Fee Association Books, London.

Haraway, DJ 1992, *Primate visions: gender, race, and nature in the world of modern science*, Verso, London.

Haraway, DJ 2006, *The companion species manifesto: dogs, people, and significant otherness*, Prickly Paradigm Press, Chicago.

Helmreich, S 1998, *Silicon second nature: culturing artificial life in a digital world*, University of California Press, Berkeley.

Ingold, T 1996, 'General introduction' in *Key debates in anthropology*, ed. T Ingold, Routledge, London.

Krantz, SL 2002, *Refuting Peter Singer's ethical theory: the importance of human dignity*, Praeger, Westport, Conn.

Kurzweil, R 2000, *The Age of spiritual machines*, Penguin Books, New York.

Kurzweil, R 2013, *How to create a mind: the secret of human thought revealed*, Duckworth Overlook, London.

Lakoff G 1987, *Women, fire, and dangerous things: what categories reveal about the mind*, University of Chicago Press, Chicago.

Lakoff, G & Johnson, M 1980, *Metaphors we live by*, University Chicago Press, Chicago.

Lakoff, G & Johnson, M 1999, *Philosophy in the flesh: the embodied mind and its challenge to western thought*, Basic Books, New York.

Lin, P, Abney, K & Bekey, GA 2011, 'Introduction' in *Robot ethics: the ethical and social implications of robotics*, The MIT Press, Cambridge, Mass.

Malik, K 2000, *Man, beast and zombies: what science can and can't tell us about human nature*, Weidenfeld and Nicolson, London.

Minsky, M 1967, Computation: finite and infinite machines, Prentice-Hall, Englewood Cliffs, N.J.

Minsky, M 2007, *The emotion machine: commonsense thinking, artificial intelligence, and the future of the human mind*, Simon & Schuster, New York.

Mol, A 2002, *The body multiple*, Duke University Press, Durham.

Moravec, H 1998, *ROBOT: mere machine to transcendent mind*, Oxford University Press, Oxford.

Nitchie, EB 1913, *Lip-reading principles and practise: a hand-book for teachers and for self instruction*, Frederick A. Stokes Company, New York.

Nitchie, EH 1950, *New lessons in lip reading*, Lippincott, Philadelphia.

Strathern, P 1997, *Turing and the computer*, Arrow, London.

Thompson, A 1996, 'Introduction' in *Julien Offra de la Mettre: machine man and other writings*, ed. and trans. A Thompson, Cambridge University Press, Cambridge, pp. viii–xxv.

Turing, AM 1950, 'Computing machinery and intelligence', *Mind*, vol. 59, no. 236, pp. 433–460.

Turkle, S 2011, *Alone together: why we expect more from technology and less from each other*, Basic Books, New York.

Walker, FR 1999, *Lip reading*, Diamond Twig, Newcastle upon Tyne.

Weiner N 1961, *Cybernetics: or control and communication in animal and the machine*, MIT Press, Cambridge, Mass.

Wolfe, A 1991, 'Mind, self, society, and computers: artificial intelligence and the sociology of mind', *American Journal of Sociology*, vol. 96, no. 5, pp. 1073–1096.

Films Cited

2001: A Space Odyssey 1968, dir. Stanley Kubrick.

3   Social Robots

AI: Artificial Intelligence (2001) is a story about the relationship between a robot child and a human mother. This Stanley Kubrick and Steven Spielberg collaboration is one that extends the themes in Kubrick's earlier work, *2001: A Space Odyssey* (1968) by inviting the viewer to think about intelligence and relationality. The film also follows a lineage of Spielberg's cinematic portrayals of intelligent nonhuman life forms (reflected in his interest in alien narratives). The nineteenth century tale of Pinocchio provides the backdrop to the screen fiction. Pinocchio is an artificial entity seeking "life" and "to be a real boy", just as the robot character in the film. In this narrative, Mister Geppetto's craftwork Pinocchio is made of wood with screws setting his joints in place. David, the futuristic AI creation of the Mecha Corporation, is an updated technological reworking of the Pinocchio narrative and is an AI: a computer-programmed entity made from metal, wires, circuits, and transistors and covered in synthetic, human-like skin. David is a prototype robot, designed as a companion robot for couples lacking or wanting a child. The themes of companionship and the longing for a child remain central to both. Pinocchio's maker, Geppetto, is a master artisan and crafts a wooden puppet boy. Hence, both examples indicate fecundity in male acts of artifactual creation that they, too, can rival the biological capacities of women to reproduce. This theme is also reproduced in the fiction story *AI: Artificial Intelligence*. The CEO of the Mecha Corporation is a Geppetto in another form. He, too, wants to reproduce artificially, namely robots.

AI: Artificial Intelligence is about the longing of a child robot for a mother. David is imprinted with the following code: 'Cirrus. Socrates. Particle. Decibel. Hurricane. Tulip. Monica. David. Monica' (*AI: Artificial Intelligence* 2001), and on speaking this code, David is forever imprinted with feelings of love for her. This imprinting harkens back to the Harry Harlow experiments on young animals and Konrad Lorenz, who carried out a number experiments on geese and ducks to show that an animal would attach to the first thing it saw, sometimes Lorenz himself (Pearce 2009). Returning to the story, Monica cannot return the intensity of love for robot David. She sees him as a machine and fearing that he may hurt her other child Martin, she abandons him in the woods rather than returning him to Mecha

Corporation to be destroyed. The story does not end there, however, as David then ventures on a quest to be reunited with his mother. He goes on a perilous quest, meeting other robots along the way, such as Gigolo Jo (Jude Law), a lover robot. That the film's main robot characters are both concerned with providing human love, intimacy and contact is telling.

As an anthropologist, the film raises important questions for me regarding human-machine relationality and attachment. Is it really possible that machines could become viable alternatives to other persons? Could robots become our lovers, children, therapists and friends? While in the lab at MIT, I came across the term "companion species". Social scientists also provide a framework for reflecting on multiple significant othernesses. Donna Haraway's (2003) book *The Companion Species Manifesto* is an argument for human-nonhuman relationality. That I should find a coincidence of term and project was revealing, showing a bridge between robotic science and anthropological theorizing of social-relationality. The extension of relationality to nonhumans (dogs, robots, machines) was a trend simultaneously occurring in anthropology and robotics.

In labs at MIT and beyond, robots are reimagined by researchers as childlike, companions, friends and a different kind of special altogether. There was a value placed on the theme of relationality, and this was structured mainly around designing machines that could mimic, imitate and resemble the caregiver-child relationship. That children, their development, capacities, and ability to bond and be supported by adults became interesting questions for robotic scientists, who reflected on these arrangements for the development of their own machines. Since conducting my fieldwork at MIT during 2003–2005, the child robot is a feature of many labs worldwide, including former and current ones: socially interactive robots Kismet and Mertz (MIT), the robot Bandit (University of Southern California, US), KASPAR (University of Hertfordshire, UK), RobotCub (various labs in Europe), ASIMO (Honda, Japan), and Biomimetic Baby or CB2 (Japan). Even if the robot is not intentionally designed to look like an infant or childlike, many still incorporate notions of child developmental psychology and epigenetic robotics.

ROBOTIC CHILDHOOD

The former Artificial and Computer Intelligence Laboratory (now the Computer Science and Artificial Intelligence Laboratory) is located on the MIT campus in a now defunct building called NE43. NE43 is composed of multiple floors that host various research groups in AI and robotics. The robotics lab was situated on the top of floor of NE43. At MIT each building has a code, and as one researcher explained to me, you can pinpoint any location at MIT in code form. The use of codes to navigate the built environment at MIT is part of its etic and reinforces the notion that MIT is a site of logical, impersonal rationality, where codes and numbers shape practices. NE43 no longer exists as the site of the robotics lab. The building was built circa 1960 and hosted the first Artificial Intelligence Laboratory, but in early 2004, the occupants of NE43 moved into the Ray and Maria Stata Center (Building 32). The building is something of a cause célèbre on campus, as it was designed by Frank Gehry and was informed by radical anti-dualist building design practices (Gilbert-Rolf & Gehry 2002; Wright & Thompson 2004; Mitchell 2007).

My ethnography takes place in NE43 and in the Ray and Maria Stata Center. I came across my first robots at NE43. As you walk along the corridors of the ninth floor of NE43, you will get a sense of a building in decline. The mosaic walls are filled with edges of posters, hanging partially with the scotch tape that once held them intact. New posters of new robotic projects have replaced these forgotten images, layer upon layer of different projects showcased once at some conference or workshop, now left here to decorate the blandly-colored walls. The lab area of NE43 is chaotic—wires, leads, cables, circuit boards, transistors, screens, data banks fill the space; a "disorderly chaos" best describes the place. As you exit an elevator on the ninth floor of NE43, once you get passed the secured doors, you will enter a different world. The floor was a mixture of open planned spaces and closed hermetic office seals. Most researchers and faculty on the floor kept their doors open. As you move from the part of the floor where the offices are located, you will then find on the other half of the floor the robot labs. It is here that I saw my first robot, Marius.

Marius was the brainchild of multiple researchers. The professor of the lab was inspired to build a humanoid robot after watching a rerun of the film *2001: A Space Odyssey* with his students in 2001. Marius was around 5 feet, 8 inches high. It was very machine-like, as its wires, cameras were exposed and left for a viewer to see, apart from a tight Velcro bodice that covered its torso area. Marius had different body parts at different times: one arm, then two arms, one hand, two hands. One thing was certain: it was fixed to the floor and immobile. A score of cameras and cables were lined behind Marius, processing all the data from its complexity.

Though Marius looked slightly like the Terminator robot from the films of the same name, it was surprisingly not thought of so negatively. *The Terminator* films and other films like them did underscore the design philosophy of Marius. If anything, Marius is a counter-cultural robot and may look like a relative of the Terminator, but I was encouraged to see Marius as something more innocent. Surrounding Marius, children's toys such as puzzles, teddy bears and books are not items you would expect to find in robotic labs, but these are the kinds of objects you will see commonly. The robotic scientists worked on Marius as though it were a small child.

In the making of robots, the researchers have developed an entire philosophy about the ways in which the robot should be constructed and rules for interactions with it. These models are drawn from child developmental psychology. In particular, robotic scientists are developing childlike machines; that is, they ascribe multiple childlike qualities to their machines: facial features, child cognitive models and socializing techniques of nurturing and play. In some cases, they imagine themselves as their parents, as noted by Cynthia Breazeal (2002), a notable social robotics expert, who was responsible for designing the social robot Kismet (with colleagues at the former AI lab at MIT in the late 1990s and early 2000s) saw herself as "mother" to the machine. Robotic scientists believe that building robots like children is effective. If the interlocutor relates and accepts the robot as a child, the adult is more likely to provide this supporting or "scaffolding" role to help the robot learn about the environment, in which case there is an affective bond that is encouraged between adult and robot child. As robotic scientist Brian Scassellati explains:

> Humans are not born with complete reasoning systems, complete motor systems, or even complete sensory systems. Instead, they undergo a process of development where they are able to perform more difficult tasks in more complex environments en route to the adult state.
>
> (2001, p. 29)

Why should humans be interested in supporting the development of machines?

In making robots physically resemble children, robotic scientists aim to challenge cultural images of robots as threatening. Films such as *The Terminator* tetralogy (1984, 1991, 2003 and 2004) and *The Matrix* trilogy (1999, 2003, 2003) acted as the backdrop. In these films, robots and AI machines are threatening and intent on destroying or controlling humans as they fight for supremacy. In making robots as children, these research scientists believe they can encourage greater intimacy and affection between humans and machines. Robotic scientists are conscious of the cultural perceptions of robots as competitors to humans, or threats, ones that may turn on their human creators. In this sense, then, robots as children appear less threatening; they evoke alternative feelings, and they attract a degree of interest and even compassion. In other ways, too, a robot that appears like an infant can activate a whole different kind of social architecture amongst humans who interact with them. Adults interact with human infants in particular kinds of ways. The lack of language or mobility in a child does not prevent people from engaging with them (Bee 1975). Research scientists want to engage a spontaneous outcome of child-adult arrangements by providing a comforting engagement between robot child and adult. Let us explore this issue in relation to the robot Radius.

Radius is a sociable robot and was put in public spaces around MIT to encourage interactions with passers-by. Radius is designed explicitly to look childlike. Radius has a grey face, large eyes, eyebrows and a mouth and has been compared to Casper the ghost. The interactions that took place between Radius and the (mainly) adults that would interact with it were stifled and required considerable effort on their part. As Radius was placed in public corridors, there were many opportunities to observe how random individuals would interact with the machine. Some passersby recorded these interactions and put them on sites such as YouTube. In one such video, a male and female of European-descent (judging from the accents of the speakers) try to engage Radius in a conversation. The man and woman try endlessly to understand Radius's chaotic behaviors. 'Please look into my eyes', it would repeat over and over again. The couple would then try to engage Radius by looking into its eyes, only to see it chaotically shift from one setting to another, continuing its requests: 'Please look at me directly', 'Please look into my eyes', 'You are too far away', Radius would cry. When I watch this video, I start to feel sympathy for Radius and want the robot to make a connection with the adults. The video lasts for a few minutes, and the adults try over and over again to make a connection with the robotic machine, but no such connection is really made despite the visitors' and the robot's attempts. Their interactive acts are out of sync.

Robotic scientists use models and theories of child development when thinking about, planning and implementing research ideas in robots. In the field of robotics, there is a whole sub-field of research that deals specifically with this area of work, namely epigenetic robotics, that is well enough established to host annual conferences and possess a research journal. Epigenetics is the study of robots using child-based models of psychology and infant development, known in the field as epigenetic robotics. It was a concept introduced into the field by Jean Piaget and was used to refer to the 'the interaction between the organism and the environment, rather than by genes' (Zlatev & Balkenius 2001, p. 1). Piaget described intelligence in the following way: '. . . I would define intelligence as a form of equilibration, toward which all cognitive functions lead' (1962, p. 110).

Piaget distinguished four periods of the development of intelligence, 'first, the sensori-motor period before the appearance of language; second, the period from about two to seven which precedes real operations; third, the period from seven to 12 years of age, a period of concrete operations (which refers to concrete objects); and finally after 12 years of age, the period of formal operations, or propositional operations' (1962, p. 110). Robotic scientists are working to develop robots in the first two stages of child development. Piaget's importance of sensorimotor development is interconnected with embodiment, as children's spatial, causal and emotional activities are driven by their interactions with the world of people and things (Gibbs 2006, p. 8). The sensorimotor stage is most important because it is the primary stage where children acquire control over their own bodies. Robots are often compared to children at this stage as they, too, lack control over their bodies. As Piaget explains, 'Before language develops, there is behavior we can call intelligent . . . when a baby of 12 months or more wants an object which is too far from him, but which rests on a carpet or blanket, and he pulls it to get to the object, this behavior is an act of intelligence' (1962, p. 110). Piaget noted how children use intermediaries to get what they need, and emphasized the importance of the sensorimotor stage as the time when the child will develop ideas of space, of time and of causality (1962, p. 112). Robotic science is a blend of child developmental theory with embodiment and cognitive science.

In robotic literature, journals feature childlike images and refer to child-adult relations as the analogical platform for robot-human relations. The paper 'Learning to Grasp with Parental Scaffolding', in which the authors (Ugur et al. 2011), propose an experiment where a robot's attempts at grasping are improved by a caregiver's guidance. Scaffolding is a term taken from developmental psychology in which support from an (adult) caregiver provides assistance to help a child with cognitive development (Ugur et al. 2011), and is frequently cited as a concept in robotic science (Breazeal & Scassellati 2000; Breazeal & Velásquez 1998). Infants are unable to control their bodies and require continual assistance from adults for many years (Bee 1975). Typically developing infants possess the fundamentals of sight, hearing, taste and touch, but these are shaped and cultivated when the child interacts with adults and their environment (Bee 1975, pp. 73–76). Social scaffolding between adults and children is an architectural structure mimicked by robotic scientists when developing social robotic systems.

THE HUMAN BOND

In reflecting on the ways in which robots are imagined to fill gaps in human social relationships, as companions, lovers, children and therapeutic agents, it is important to understand how humans make each other and what theories have been offered to explain this remarkable feeling of connection one human feels for another. How do humans form bonds and attachments to each other? A child's development is relational, as the child would not survive without a caregiver. A child's development is multiple involving many parts: cognitive, sensorimotor, feeling, agentive and sensuous developments simultaneously occur (Stawarska 2009). As a child's development is always relational, the relation is of paramount importance. The first stage of attachment will be between the newborn infant and caregiver(s). Piaget's theories of child psychology explored the cognitive, sensorimotor and agentive aspects of infant and child development (Piaget 1930; Piaget 1929), and John Bowlby (1981) explored the feeling component of the attachment. According to Bowlby, a child will learn about attachments in infancy, and these become the models that the child takes into their relations with others outside the family and/or caregiving situation. As Hazam and Campa (2013) argue, 'There is overwhelming evidence that the quality of relationships with primary caregivers during infancy can have lasting effects on how one subsequently relates to others' (p. 2). Psychological evidence is drawn from exploring disordered or wounded attachment patterns and then developing strategies or techniques to repair attachment wounds (Bowlby 1981; Ainsworth 1967; Ainsworth et al. 1978; Hazam & Campa 2013; Dykas & Casidy 2013; Brisch 2012; Pearce 2009). There are some children who find difficulty in forming bonds and the term to describe this is autism, where there are considerable difficulties in social-interaction and reading cues (Baron-Cohen 1995; Baron-Cohen 2003).

The etymology of the term "bond" is from the twelfth-century English "bind", which means ties and binds. This can be interpreted as one entity that is connected to another. One cannot see a bond nor touch it, but bonding is an important aspect of human existence; knowing what it is and how to create one, however, is a controversial topic (Marrone 1998). Philosopher Martin Buber (1937) proposed a philosophy of relations. He described these relations in two ways: I-It and I-Thou. Buber believed the I-It relation was based on experience and partiality, of taking something from the other, but the I-Thou relation occurred with the whole of the being and was dialogical. For Buber 'All reliving living is meeting' (1937, p. 11) and in the encounter, in the exchange between one and the other (human, nature, spirit). Buber emphasized the importance of presence in relation to the other 'Love does not cling to the *I* in such a way as to have the *Thou* only for its "content," it's object; but love is between *I and Thou*' (1937, pp. 14–15).

Child psychologist John Bowlby controversially proposed that childhood attachments to a significant, responsive and loving caregiver are crucial for the healthy development of children. Adult attachment patterns may seem unrelated to their youthful counterparts, but Bowlby argued that one directly informed and developed the other (1981). Children who do not have access to a stable and consistently loving caregiver will have difficulties forming attachments in adulthood (Hazan & Campa 2013). Recent research has begun to explore how early attachment patterns shape adult styles of relating (Pearce 2009).

Why is the child-carer bond so fundamental in shaping human relationships? Researchers suggest it is because the human child goes through a considerable period of dependency on an adult for their development (Ainsworth et al. 1978; Dykas & Cassidy 2013). Children's emotional, survival and cognitive needs cannot be met independently of adult support. Love is an important ingredient of this bond because love acts as a kind of glue that binds the caregiver and child (Bowlby 1981). Bowlby's work was important and led to pioneering work in the area of childhood attachment, but it also called into question the notion that a child-carer bond was an automatic outcome of close proximity. For attachment bonds to form, a special kind of connection was required between carer and child. Mary Ainsworth, in her famously titled 'Strange Situation Experiment', explored these issues further (Ainsworth et al. 1978). In the experiment, children were observed reacting to a parent leaving and returning to a room, and she observed feelings of distress, indifference and pleasure displayed by the child. She proposed from the experiment three main types of attachment bonds—secure, avoidant and ambivalent/anxious. Secure attachment is characterized by pleasure and comfort in relationships with others, but when children have avoidant and ambivalent/anxious attachment styles, they can find relationships confusing and frightening (Brisch 2012). Some researchers have challenged the notion of childhood as a special kind of existence, drawing attention to Philippe Ariès and his famous work *Centuries of Childhood* (1973). Modern childhood is seen as a distinct phase of development, and children are considered radically different from adults.

Much of the research done in the area of childhood attachment is from studies of problematic attachment patterns (in children and adults). Children brought up in homes where violence, sexual abuse, illness and neglect are a regular feature of life are significantly more likely to grow up with interpersonal and psychological problems (Bowlby 1981). Moreover, early trauma in children has lifelong consequences, with adults suffering from a higher prevalence of mental health issues, cognitive difficulties and difficulties in forming pleasurable and safe adult relationships (Pearce 2009; Reis 2013). The bond then, is not just an automatic outcome of proximity, but a particular kind of relationship that is cultivated between carer-child, and this style then flows into adulthood. This view of attachment as an outcome of loving bonds initially formed between carer and child, and then extended into other relationships, has significant consequences in reflecting on contemporary models of relationality and sociality. Robotic scientists draw our attention to the importance of childhood in the development of humans by mimicking the child-carer relationship and by seeing robotic machines as children.

SOCIAL RELATIONAL ROBOTS

Robotic researchers are interested, then, in the social dynamics that take place when social bodies interact. They focus on aspects of the body that emphasize communicative interaction, such as the importance of specific bodily parts; among these are eyes and general body shape that mimic human bodies and hands.

The robots Radius and Marius were built with these types of bodily communicative specifications in mind. The construction of the body is one step towards modeling a socially interactive body. Human bodies carry within them information about communication. That is to say, they are given a familiar socially interactive framework within which to operate. A human body form carries with it certain expectations about both its behavior and performance. The simple behaviors of nodding the head, pointing a finger, looking in a similar direction, orienting body direction, smiling, frowning and tilting the head to one side, pulling back the head and raising the eyebrows all demonstrate cooperation, shared interest, approval or dislike, surprise, and curiosity. In the absence of any sophisticated behaviors, many types of "interior" states or "relational" states can be communicated by a robot's behavior. Researchers at MIT went one step further than this mere puppetry by attempting to engage a human in a communicative behavior-based dialogue with a machine. This was most notably done by the robot Kismet, the creation of Cynthia Breazeal and several of her colleagues at the MIT AI Lab.

The robot Kismet was an example of a sociable robot, and the first of its kind designed by the former AI Lab at MIT. Kismet was a research platform and had a head with facial movements that mimicked human expressions such as being angry, bored and happy. It had a limited number of facial movements that enabled it to mimic these basic human emotions. The inspiration for Kismet was drawn from Rodney Brooks' model of behavior-based robotics. Kismet was to behave in such a manner that it could communicate emotion through facial expressions and bodily movements (the bodily movements confined to the head). If a robot is to perform sociable interactions, it needs the same physical parts that humans also use in the same interaction. Kismet mimicked human behaviors, yet the sociable repertoire of Kismet was not exclusively human—it also mimicked common animal behavior, such as showing fear by withdrawing. Kismet's technology was simple, yet extremely effective. Kismet had a series of drives that regulated its patterns of interactions with human interlocutors. These "drives" were randomly ordered and, by altering its behaviors, Kismet may appear happy or sad or interested. As the drives were randomly ordered, Kismet could move from one drive to another. The makers added another element to Kismet: they made Kismet responsive by deploying senses to help it choose appropriate drives/emotions for a situation. If Kismet was interacting with an interlocutor, the behavior of the interlocutor could influence the sociable-emotional responses of Kismet. This depended on one's proximity to and interaction with Kismet for it to alter its emotional arrangements. It would indicate distress if one got too close and indicate boredom if one was too far and out of its visual range. Rodney Brooks had this to say about Kismet:

> Kismet is the world's first robot that is truly sociable, in that it can interact with people on an *equal basis*, and which people accept as a humanoid creature. They make eye contact with it; it makes eye contact with them. People read its mood from the intonation in its speech, and Kismet reads the mood of people from the intonation in their speech. Kismet and any person who comes close to it fall into a natural social interaction. People talk to it, gesture to it, and act socially with it. Kismet talks to people, gestures to them, and acts socially with them. People, at least for a while, treat Kismet as another being. Kismet is alive. Or may as well be. People treat it that way.
>
> (2002, p. 65; my emphasis)

Kismet allowed researchers to think about sociable interaction with machines. My interlocutors were involved in the making of Kismet, but

I only had direct contact with Kismet behind the glass of a museum case and from watching videos of Kismet in action. In order to give Kismet its animate and life-like qualities, Breazeal imaginatively enrolled the work of Disney animators, who are adept at creating believable characters that can trigger specific emotions and thoughts in spectators (Breazeal 2002, p. 163). Animation techniques are used in particular ways in the making of sociable robots. The eyes are very important in this respect. Animators use eyes to communicate emotional expression and often exaggerate the size of eyes in cartoons.

The researchers identify body part with activity. In giving robots sociably expressive characteristics, do interlocutors feel involved in relations with them in new ways? Did the human-like and sociable qualities of Kismet convince those who interacted with it that it was more than a machine? Or did Kismet's makers "trick" those it interacted with into thinking it was "living" or "emotional"? More importantly, however: does it matter?

If robots can carry out convincing interactions with human beings, then does this alter the status of these kinds of robots? Breazeal raises the following questions:

> When is a machine no longer just a machine, but an intelligent and 'living' entity that merits the same respect and consideration given to its biological counterparts? How will society treat a socially intelligent artifact that is not human but nonetheless seems to be a person? How we ultimately treat these machines, whether or not we grant them the status of personhood, will reflect upon ourselves and on society.
>
> (2002, p. 240)

The interest in relational computational machines began earlier. Josef Weizenbaum created the computer program ELIZA in 1963 as a therapeutic aid. ELIZA works by generating questions in response to questions. The ease with which (some) people would interact with ELIZA led Weizenbaum to ethically assess his own morality as a computer programmer. He was even more disturbed when Carl Sagan, a science visionary, speculated on the possibility of replacing therapists with machines (1984, p. 5). Weizenbaum wrote disturbingly of these therapeutic aspirations: 'What can the psychiatrist's image of his patient be when he sees himself, as therapist, not as an engaged human being acting as a healer, but as an information processor, following rules, etc?' (cited in Brooks 2002, p. 167). Weizenbaum's ELIZA illustrates not only how humans and machines become integrated via relational machines, but also how specialized practices (psychoanalysis or cognitive behavioral therapy) are subject to rules that can be modeled and transcribed into computer programs. Weizenbaum reacted fiercely to the popularity of his program and rejected claims from the psychiatric profession that the program could be used as the basis of counseling and psychiatry. Wizenbaum might have been shocked to discover that there are several different types of therapeutic counseling programs that are available today. For example, it is possible to purchase for £110 the 'Blues Begone' program, a computer-based therapeutic counseling system. Robots are now imagined as therapeutic agents to help children with autism develop social skills and act as substitute therapists.

Sociable robotics developed as a sub-field of humanoid robotics (and, arguably, out of the insights of human-machine interaction) in the early 1990s. When I first entered the lab in May 2003, I was impressed to hear that the robotic scientists had plans to make humanoid robots, some of these being social robots. On hearing this, I began to imagine these robots were of the variety I had expected in childhood, that they were creating a new, sophisticated type of domestic robot. The *Jetsons*, an animation from the 1970s of a romanticized 1950s life, featured space travel, high-tech life and, of course, Rosie the Robot, a domestic servant robot. While Rosie was sophisticated, her primary purpose was as a helper to the family's domestic chores. This transformation of perspective of robots is arguably the most critical re-evaluation of the robot's purpose and place in society. The cultural imagination of robots has significantly shifted to allow for types of social constructions where the machine is implicitly imagined as a potential possessor of personhood, with relationality as the ontological dynamo. In relationality theorizing, entities acquire transformative properties in relations and their ontological status as machines, inert entities or functional objects, with specific kinds of affordances given by different cosmological orderings.

These themes are illustrated through the relational studies of Sherry Turkle (1984), arguably one of the pioneers of human-machine relationalities. Turkle developed a framework with which to understand objects that evoked specific kinds of affective feelings, describing them as a 'relational artifact'. A relational artefact is any object that one has a relation with—be it a Furbie, robot or musical instrument. One such relational artefact is the Tamagotchi, created in 1996 by Aki Maita who came to be known as the 'Mother of the Tamagotchi'. The Tamagotchi is a hand-sized virtual pet that its users have to feed and tend to keep alive. If users do not follow commands from the Tamagotchi, the virtual creature will cease to function. In this context, the Tamagotchi revealed the extent to which its users were obligated to a machine to keep it virtually living. The Tamagotchi, for instance "dies" when unfed—that is, if someone does not attend to the Tamagotchi on a regular basis, the system closes down. The first round of Tamagotchis terminated completely on the 'death' of the creature, but later versions allowed the user to reset the functions. There are examples of young owners taking these 'dead' objects to Tamagotchi burial grounds. Is the extension of death-rights to virtual pets an indication of the changing boundaries between human and machine?

Tamagotchis simulate hunger, happiness and death; they give the user a common framework within which to engage and interpret the behavior of the virtual creature by simulating hunger, play, death and emotions. Robots and virtual creatures stimulate the same feelings in people. What is perhaps more surprising in the field of robotic toys (and now many lab-based research platforms) is that robots are designed to recognize other robots. On recognition, a robot can activate a script that allows it to interact with another of its kind. The now discontinued AIBO robot can recognize its owner, as well as other AIBO dogs. If an AIBO comes into contact with another AIBO "dog", it will bark and "play" with it. The fact that AIBOs recognize their owners, as well as other AIBOs, is a curious development. In effect, such an activity indicates that its "social" needs can be met by humans and other machines.

Theories of relationality privilege the site of interaction in the dynamic and are the outcome of anti-essentialist theorizing. If essentialism is the kind of specific quality a person or thing possesses, then the role of the essentialist is to categorize and label those qualities. Anti-essentialism asserts that there is nothing particularly essential about an entity, but the entity itself is created in the relation. This is a new kind of relations theory, where relations between entities become paramount factors in the agency of the creature.

Robot scientist Scassellati (2001) proposes a theory of mind (ToM) for a robot and outlines what is needed for robots to act as embodied minds:

> *Attribution of Animacy:* The ability to distinguish between animate and inanimate objects on the basis of the spatial and temporal properties of their movement.
> *Joint Attention:* The ability to direct attention to the same object to which someone else is attending.
> *Attribution of Intent:* The ability to describe the movement of pairs of objects in terms of simple intentional states such as desire or fear.
>
> (2001, p. 17)

An important component of this anthropomorphism process is applying a ToM to the robot (Scassellati 2001, p. 15).

In sociable robotics, the emphasis is put on the external and visible behavior of the robot. Robotic scientists locate intelligence and sociability in the visible realm. The physical body is combined with behaviors to involve humans in a relation. Robotic scientists mimetically reproduce the analogous characteristics of a human in a similar situation. Robots communicate by their bodies and emotions. Interlocutors with robots judge their "internal" states by what is occurring in the external realm.

A parallel might be drawn with another theory, this time in the social sciences: that of actor-network theory (ANT), which aims to examine objects in terms of their effects on humans. As Strathern explains ANT, 'Actor Network theory pays attention to the way in which social relations, and their self-empowering manifestation in human skills, summon the properties of, and thus enroll the effectiveness of, artefacts and techniques regardless of whether these are (in the Euro-American ordinary language sense) person, things, animals or, for that matter events' (1999, p. 17). Strathern explains Gell's work on art: 'Euro-Americans often think agency inappropriately personified when it is applied to inanimate entities, but that is because we link agency with will or intention' (1999, p. 17). Strathern continues: 'as far as efficacy on others is concerned, one may thus see an art object in the same way as one may see a person. It embodies capacities' (1999, p. 17).

Actor-network analysis begins from this visible realm, regardless of the differences in entities involved. In fact, it sees the type of entity predominantly in relation to its performance. Thus the realm of the visible becomes the site of activity. Actor-network theory has also posited a radical anti-essentialist position. As Law explains, 'actor-network is, has been, a semiotic machine for waging war on essential differences. It has insisted on the performative character of relations and the objects constituted in those relations' (Law 1999, p. 7). Latour's view of 'interobjectivity' expresses this: '. . . [there is] nothing especially human, in a local interobjective encounter' (Latour 1999, p. 18).

I am interested in the concept of interobjectivity and performativity to describe those relations, as robotic scientists too rely on the performative and interobjective aspects of the making of sociable robots. In this sense, potentially different entities become symmetrical. Latour writes that 'Symmetry is defined by what is conserved through transformations. In the symmetry between humans and nonhumans, I keep constant the series of competencies, of properties, that agents are able to swap by overlapping with one another' (1999, p. 182). If entities—human and nonhuman—derive meanings in relation to each other, then the meaning that they have is dependent on the relation that methodologically situates the visible realm as the site of that contextual assemblage.

THE NONHUMAN IS HUMANIZED

In general, humans seem to possess a spontaneous cognitive capacity to project social agency and narratives on nonhuman phenomena, though this is difficult in some children and adults with autism (Kanner 1943; Baron-Cohen 1995). In a famous study by Fritz Heider and Marianne Simmel entitled 'An Experimental Study of Apparent Behavior' (1944), the authors showed a simple animation to subjects and asked the subjects to describe what they believed was happening. The animation featured a black rectangle with a corner that would open and close, and three interacting geometrical figures: a large triangle, a small triangle, and a disc. The geometrical shapes moved randomly in the animation and the subjects were invited to comment on the apparent behavior of the objects. Nearly all the participants in the experiment gave the animation a social narrative form. There were no humanoid figures in the animation, but the movement of the shapes and their dynamics was enough to provoke a socially-inspired narrative. Many of the subjects projected a familial narrative; they saw the smaller shapes as children and the larger shapes as parents or adults. Moreover, many participants interpreted the actions of the larger shape as one of cruelty, acting unfavorably to the smaller shape. Heider and Simmel claim there was no consciously intended narrative. If one were to look logically at the images, they were just shapes moving around on a screen, but humans see social narratives everywhere.

Stewart Guthrie's book *Faces in the Clouds* (1995) is an apt title for indicating the contents of the book, which is the human cognitive disposition to anthropomorphize, animate and attribute intentions to nonhuman things. Guthrie is interested in how human-centered-perceptual cognition is applied to nonhuman animals and things, showing a natural-intuitive disposition of cognition. The attribution of cognitive and social states beyond humans shows that the imagination of humans is open to attributing states to nonhuman animals and things. Here, a paradox presents itself in humanoid robotics because the more the object is humanlike, the more parity is expected between the object and what it is meant to represent. In our case, the more humanlike the machine, the greater the expectation that it will behave as such.

Guthrie explores anthropomorphism in relation to religion and other subjects, including art, philosophy and science, but I am interested in the concept of anthropomorphism in relation to humanoid robots. The connection between religion and robots has been explored by Vidal (2007). Anthropomorphism is a concept with multiple meanings, and has an importance in many fields—religion, magic and animal behavioral studies to name a few (de Waal 1996, p. 7). Though robots are made in labs by experts in the specialist fields of electrical and mechanical engineering and computer science, robots are not merely technological objects; rather, they are cultural objects with meanings that are not exclusively controlled by their makers.

Despite the propensity of humans to attribute human-like qualities to nonhumans, anthropomorphism is a problematic concept in anthropology for some (Haraway 1992, 2003), and in other disciplines, as well (de Waal 1996), because it locates humans as the main agent in relations with materialities and nonhumans. But who else would be the main agent? There is a point, anthropologically, that seeing animal lifeworlds from the perspective of humans can confuse and elide the meanings that underscore these different existences (de Waal 1996), such that nonhumans merely become receptacles for normative categories of race, sex, gender and class, as was illustrated in our case above in the Heider and Simmel (1944) case, where disciplinarian parenting styles become projected onto the images. Haraway challenges this perception fiercely in her studies of simians, cyborgs and dogs (1992, 2003), but then extends it by mapping histories, genealogies and coexistence onto animals. The emphasis on hybridity (Latour 1993; Law 2001) and relational nature/culture mixtures (Haraway 2003), in my view, does not adequately explain the multiple ways in which anthropomorphism operates in everyday Euro-Americans' interactions with nonhuman animals, machines and things. Moreover, the emphasis by some scholars on hybridity and relationalities between persons and things

diminishes the human imagination in these processes. The human spectator plays a crucial role in the way in which anthropomorphic entities are
configured. While humans may interact with things, and such things trigger thoughts, feelings and behaviors, these practices are mediated through
human socialities (Gell 1998; Guthrie 1995). Therefore, anthropomorphism is not merely a frame within which to understand human interactions with technologies that have human qualities, but it is also a means
by which to rethink the importance of human sociality in underscoring
interactions with nonhumans.

Anthropomorphism can refer to several different types of processes of attributing human characteristics to nonhuman animals and things. As a concept, it
has been discussed widely in relation to animals (Silverman 1997; Moynihan
1997; de Waal 1996). For Guthrie (1995), anthropomorphism is a cognitive-
perceptual process whereby humans 'guess' about the states of others, or attribute a theory of mind (ToM) to others, including other humans. In studies of
autism, the failure to attribute a ToM to other people can result in social and
communication difficulties with animals (Baron-Cohen 1995). A characteristic
attribute of children with autism is said to be a difficulty in grasping metaphorical meanings (that one thing can stand in place for another) resulting from
the different (not absent) sociality that is experienced by individuals with the
condition (Ochs and Solomon 2010). Humans use a ToM to understand other
people's thoughts, intentions and behaviors, as well as graft their sociality onto
nonhumans and see faces in clouds (Guthrie 1995).

Roboticists deliberately exploit these cognitive-perceptual intuitions of
human beings in the way they make and galvanize support for their robotic
creatures. To put it another way, robotic scientists take the anthropomorphic characteristics of children and import them into the machines they create, while attempting to restructure the relationships between human and
robotic-machine into one between adult and child, with the adult recast as
the role of caregiver, nurturer and parent in the relation.

In this chapter I have tried to show how robots are imagined as children
and how child development models are used to model their development.
There is a relational aspect to the machine that is supported by encouraging
adults to reflect on the machine as a child. This draws together humans and
machines as particular kinds of relational and bonded entities. If machines
are now imagined to perform these roles, we must return to an understanding of the conceptualization of humans and theories of human bonds that
highlight the affective aspects of exchange.

BIBLIOGRAPHY

Ainsworth, M 1967, *Infancy in Uganda*, Johns Hopkins Press, Baltimore.
Ainsworth, M, Blehar, MC, Waters, E & Wall, S 1978, *Patterns of attachment: a
   psychological study of the strange situation*, Lawrence Erlbaum Associates, New
   Jersey.

Ariès, P 1973, *Centuries of childhood*, Penguin Books, London.

Baron-Cohen, S 1995, *Mindblindness*, MIT Press, Cambridge, Mass.

Baron-Cohen, S 2003, *The essential difference: men, women, and the extreme male brain*, Basic Books, New York.

Baron-Cohen, S 2011, *Zero-degrees of empathy: a new understanding of cruelty and kindness*, Allen Lane, London.

Bee, H 1975, *The developing child*, Harper & Row, New York.

Bowlby, J 1981, *Attachment and loss*, vol. 1, Harmondsworth, Middlesex, London.

Breazeal, C 2002, *Designing social robots*, The MIT Press, Cambridge, Mass.

Breazeal, C & Scassellati, B 2000, 'Infant-like social interactions between a robot and a human caretaker', *Adaptive Behavior*, vol. 8, no. 1, pp. 49–74.

Breazeal, C & Velásquez, J 1998, *Toward teaching a robot "infant" using emotional communicative acts*, MIT AI Lab Publications. Available from: <http://www.ai.mit.edu/ projects/ntt/projects/NTT9904–01/documents/Breazeal-Velasquez-SAB98.pdf>.

Brisch, KH 2012, *Treating attachment disorders: from theory to therapy*, 2nd edn. The Guildford Press, New York.

Brooks, R 2002, *Flesh and machines: how robots will change us*, Pantheon Books, New York.

Buber, M 1937, *I and thou*, trans. RG Smith, T. & T. Clark, Edinburgh.

de Waal, F 1996, *Good natured: the origins of right and wrong in humans and other animals*, Harvard University Press, Cambridge, Mass.

Dykas, MJ & Cassidy, J 2013, 'The first bonding experience: the basics of infant-caregiver attachment' in *Human bonding: the science of affectional ties*, eds. C Hazam & MI Campa, The Guildford Press, New York.

Gell, A 1998, *Art and agency: an anthropological theory*, Clarendon Press, Oxford.

Gibbs, RW 2006, *Embodiment and cognitive science*, Cambridge University Press, Cambridge.

Gilbert-Rolfe, J & Gehry, F 2002, *Frank Gehry: The city and music*, Routledge, New York.

Guthrie, SE 1995, *Faces in the clouds: a new theory of religion*, Oxford University Press, New York.

Haraway, DJ 1992, *Primate visions: gender, race, and nature in the world of modern science*, Verso, London.

Haraway, DJ 2006, *The companion species manifesto: dogs, people, and significant otherness*, Prickly Paradigm Press, Chicago.

Hazam, C & Campa, MI 2013, 'Introduction' in *Human bonding: the science of affectional ties*, eds. C Hazam & Campa, The Guildford Press, New York.

Heider, F & Simmel, M 1944, 'An experimental study of apparent behavior', *American Journal of Psychology*, vol. 57, no. 2, pp. 243–249.

Latour, B 1993, *We have never been modern*, trans. C Porter, Harvester Wheatsheaf, New York.

Latour, B 1999, 'On recalling ANT', in *Actor network theory and after*, eds. J Law & J Hassard, Blackwell Publishers/The Sociological Review, Oxford.

Law, J 1999, 'After ANT: complexity, naming and topology' in *Actor network theory and after*, eds. J Law & J Hassard, Blackwell Publishers/The Sociological Review, Oxford.

Law, J 2000, *Networks, relations, cyborgs: on the social study of technology*. Available from: <http://www.lancaster.ac.uk/sociology/research/publications/papers/law-networks-relations-cyborgs.pdf>. [13 November 2014].

Marrone, M 1998, *Attachment and interaction*, Jessica Kingsley Publishers, Philadelphia.

Mitchell, WJ 2007, *Imagining MIT: designing a campus for the twenty-first century*, The MIT Press, Cambridge.

Moynihan, MH 1997, 'Self-awareness, with specific references to coleoid cephalopods' in *Anthropomorphism, anecdotes, and animals*, eds. R Mitchell, NS Thompson, & HL Miles, SUNY, Albany.

Ochs, E & Solomon, O 2010. 'Autistic sociality', *Ethos*, vol. 38, no. 1, pp. 69–92.

Pearce, C 2009, *A short introduction to* attachment *and* attachment disorder, Jessica Kingsley Publishers, London.

Piaget, J 1929, *The child's conception of the world*, Kegan Paul, Trench, Trubner & Co. Ltd., London.

Piaget, J 1930, *The child's conception of causality*, Kegan Paul, Trench, Trubner & Co. Ltd., London.

Piaget, J 1962, 'The stages of the intellectual development of the child', *Bull. Menninger Clin.*, vol. 26, pp. 120–128.

Reis, HT 2013, 'Relationship well-being: the central role of the perceived partner responsiveness' in *Human bonding: the science of affectional ties*, eds. C Hazam & MI Campa, The Guildford Press, New York.

Scassellati, B 2001, *Foundation for a theory of mind for a humanoid robot*, MIT Department of Electrical Engineering and Computer Science, Cambridge, Mass.

Silverman, P 1997, 'A pragmatic approach to the Inference of Animal Mind' in *Anthropomorphism, anecdotes, and animals*, eds. R Mitchell, S Nicholas, Thompson & HL Miles, SUNY Press, Albany.

Stawarska, B 2009, *Between you and I: dialogical phenomenology*, Ohio University Press, Athens.

Strathern, M 1999, *Property, substance and effect: anthropological essays on persons and things*, Athlone Press, London.

Turkle, S 1984, *The second self: computers and the human spirit*, Granada, London.

Ugur, E, Celikkanat, H, Sahin, E, Nagai, Y & Oztop, E 2011, 'Learning to grasp with parental scaffolding', *Proceedings of the 11th IEEE-RAS International Conference on Humanoid Robots*, pp. 480–486.

Vidal, D 2007, 'Anthropomorphism or sub-anthropomorphism? An anthropological approach to gods and robots', *The Journal of the Royal Anthropological Institute*, vol. 13, no. 4, pp. 917–933.

Wright, SH & Thompson, EA 2004, 'Digital tools went into it; innovation will come out: MIT celebrates Stata Center', MIT News Office, 12 May.

Zlatev, J & Balkenius, C 2001, 'Introduction: Why "epigenetic robotics"? Proceedings of the first international workshop on epigenetic robotics: modeling cognitive development in robotic systems', *Lund University Cognitive Studies,* vol. 85, pp. 1–4. Available from: <http://www.lucs.lu.se/LUCS/085/Zlatev.Balkenius.pdf>.

Films Cited

AI: Artificial Intelligence 2001, dir. Stanley Kubrick and Steven Spielberg.

2001: A Space Odyssey 1968, dir. Stanley Kubrick.

The Matrix 1999, dir. The Wachowskis.

The Matrix Reloaded 2003, dir. The Wachowskis.

The Matrix Revolutions 2003, dir. The Wachowskis.

The Terminator 1984, dir. James Cameron.

Terminator 2: Judgment Day 1991, dir. James Cameron.

Terminator 3: Rise of the Machines 2003, dir. Jonathan Mostow.

Terminator Salvation 2004, dir. McG.

4 The Gender of the Geek

Isaac Asimov's screen adaption of *I, Robot* (2004) is set in the year 2035 when robots are ubiquitous—doing the work of people, their presence is everywhere. Robots in this future society operate according to Asimov's three rules of robotics:

> One, a robot may not hurt a human being or, through inaction, allow a human being to come to harm and must try to save a human being in danger.
> Two, a robot must obey a human being unless this goes against the first law.
> Three, a robot must save itself unless this goes against the first or second laws.
>
> (1979, p. 31)

These rules were first published in a short story, "Runaround", in the 1940s (Asimov 1979). The film's main nonhuman character is a robot, Sonny, who wants to be more than a robot; he wants rights and recognitions and feels he is more than a machine. This theme of robots desiring more than their engineers and programmers intended is a staple of many fictions of robotics and an important one when reflecting on the relations between humans and machines. How do the machines become human? What needs to happen to robots to become human?

Robots and AI systems are either cast as cold and calculating logic bots, capable of "pure" cognitive reason without emotion, or they are characterized as rational bots, desiring relationships and love for completeness. Cultural narratives of robots and machines may point to the distinctions, but AI robotics and computer scientists emphasize the similarities between humans and machines, seeing the human as a very complex machine. It is not that the human is radically distinctive from a machine, but the human *is* a complex machine. There is another aspect to this paradox, that males are closer to the state of the machine in that they are asocial and women are furthest because they are social. These ideas are mapped onto the types of persons that are attracted to AI, computing and robotic science and the ways this type of personhood is reflected back into the machines.

"NERDS" AT MIT AND BEYOND

If any visitor at MIT spends long enough in the computer, AI and robotics labs, they will surely hear someone somewhere say 'he is a nerd'. Though it is often a gendered "he" who is the nerd, this is not always the case, and at MIT there is somewhat of an unusual gendered equality in this respect. Visiting a bar one night in the Boston area, a robotic scientist drew me images of nerds and described their characteristics in great detail to me:

> They aren't too interested in what they look like. Mostly they wear sandals, shorts and a T-shirt at MIT. They are often socially awkward, too.

The Computer Science and Artificial Intelligence Laboratory (CSAIL) research scientists at MIT use the term nerd to describe each other, more prominently if they want to emphasize distinctive impersonal and antisocial qualities. According to the OED, a nerd is a 'foolish, feeble, or uninteresting person' (1991, p. 976). Beyond the OED meaning, being labeled a nerd can be both an insult (indicating a person who is socially awkward, lacking a sense of humor, personal hygiene problems and unpleasant to be around) or it can refer to a person positively (very smart, very able and expert). Often the perceived positive and negative aspects of the nerd characteristics go together in various formations. It would be categorically wrong for a reader to think this was merely a statement made by students of other students in the CSAIL environment. A day rarely passed by at MIT when I failed to hear a member of the staff, a researcher or a faculty member use these labels to describe the types of persons at MIT. These phrases and descriptions of persons were used often in and around the lab at MIT.

A place like MIT has to confront the stereotype of the nerd and/or geek enough that it feeds into the university's own activities, even if, as in this case, it is only to reject the label. If the stereotype of MIT students as nerds was not prevalent, why would the university authorities have to explicitly deal with these issues? MIT runs a charm school every year as part of the university's Independent Activities Period (IAP). It is predominantly men that attend the course, but women can attend too. This is what the organizers of the charm school said about nerds:

> Q: Some people say that MIT students are nerds and this program is designed to address this. Is this why you run the program?

> A: No! The classes we teach are designed for everyone, and we feel that we all have something to learn about etiquette, manners, communication, and personal skills. The nerd stereotype is not only misleading, it

does not begin to capture the diverse, dynamic network that we have here at MIT. The bottom line is that not everyone receives this kind of instruction, either at home or in school, so we feel they are important life skills for us to pass on to the network.

(MIT Charm School n.d.)

Classes at charm school include 'how to accessorize/dress for success, overcoming shyness, dating etiquette and small talk and attentive listening' (Charm School Frequently Asked Questions). However, the typecasting of science and technologists is not confined to my interlocutors; these typecasts are routinely depicted in popular culture and psychological texts, as well as being found in social studies of science (Helmreich 1998). In American culture, popular television programs and films reproduce these cultural images. In 2005, an American network aired *Beauty and the Geek*—a reality show where "beautiful" women are paired up with "intelligent men". The way in which geek/nerd culture reproduces gender dichotomies with attractiveness rated as a specifically female quality and intelligence as male. The *Beauty and the Beast* narrative is reproduced in these narratives of gendered female beauty and lack of intelligence and male intelligence and lack of physical attractiveness. Aside from issues of gender, nerd and geek narratives speak to issues of class, with White and Asian groups significantly represented (Kelty 2005). There is also a nerd and geek genre in film, epitomized by films such as the trilogy of the *Revenge of the Nerds* (1984–1992).

Cultural stereotypes of nerds are complicated in US culture due to an intersecting matrix of power, prestige, visions of genius and business acumen. Bill Gates, Steve Jobs and Mark Zuckerberg are all symbolic of this "Nerd World Order", where individuals who themselves are, according to reports, socially awkward and difficult, create social technologies (the laptop, cell phone, iPod) that allow the atomized consumer to reconnect with an online digital social sphere. What are nerds and geeks? The term "nerd" is taken from the book *If I Ran the Zoo* by Dr. Seuss (Theodor Seuss Geisel), which was published in 1950. Houtman and Zeitlyn explain:

'Nerd' is a term invented by Dr. Seuss in If I ran the zoo in 1950 where it represented a small comically angry-looking and unpleasant humanoid creature—'And then, just to show them, I'll sail to Ka-Troo and Bring Back an It-Kutch a Preep and a Proo a Nerkle a Nerd and Seersucker, too'. Initially popularised in the 1970s as a reference to uninteresting persons, as the information technology revolution turned playful hippies into serious businessmen, later films such as Revenge of the Nerds granted them intelligence as bespectacled, but unathletic maths student wizards (in opposition to the athletic and sportive jovial 'jock') who turn the world upside down with their wizardry.

(1996, p. 2)

Sherry Turkle (1984) identified the computer as a particular kind of relational artifact that was capable of generating intimate feelings from its user, writing: 'One can turn to the world of machines for [a] relationship . . . and the computer, reactive and interactive, offers companionship without the threat of human intimacy' (Turkle cited in Levy 2009, p. 68). Human-machine interaction explores the relations between humans and machines and uses this information mainly to develop more user-friendly technologies (Suchman 2006; 1987). Writing in 1987, Suchman explains the meanings of human-machine interaction and the trend to imbue technologies with human qualities. There is even a field of affective computing developed by Rosalind Picard of MIT's Media Lab that explores computing as an emotional sphere (Picard 1997).

It may be that nerds are poor at socializing, but this is rarely their choice and the consequences of social isolation are in some cases severe: 'Being unpopular in school makes kids miserable [say's Yahoo founder when reflecting on his school years and his label as a nerd], some of them so miserable they commit suicide' (Graham 2004, p. 2). Graham continues to explain why nerds make the choices they do:

> Much as they suffer from their unpopularity, I don't think many nerds would [sacrifice their intelligence for popularity]. To them the thought of average intelligence is unbearable. But most kids would take that deal. For half of them it would be a step up. Even for someone in the eightieth percentile (assuming, as everyone seemed to then, that intelligence is scalar), who wouldn't drop thirty points in exchange for being loved and admired by everyone? . . . Nerds serve two masters. They want to be popular, certainly, but they want even more to be smart.
>
> (2004, p. 4)

Why does excess intelligence seem to correlate with a deficiency in socializing and interpersonal skills? Stefan Helmreich found this in his own studies among artificial life researchers: 'I think scientists are usually just sort of socially disabled, so they enjoy forming systems in which they don't have to pay attention to . . . real social interactions, and things that they're not good at' (cited in Helmreich 1998, p. 69).

EXTREME SYSTEMATIZERS?

Do (male) AI robotic scientists have a distinct set of characteristics that separate them from persons in other academic disciplines? Simon Baron-Cohen, author of *The Essential Difference* (2003) and *Mindblindness* (1995), believes so, arguing that scientific and technological professions require specific kinds of skills such as an attention to detail and an obsessive focus on finding patterns and designing systems—psychological characteristics also found in excess in children and adults with autism spectrum conditions (ASC). In *The Essential Difference* (2003), Baron-Cohen designed a questionnaire that asked men and women to report their skills in empathizing (the Empathy Quotient—EQ) and in systematizing (the Systematizing Quotient—SQ). A typical question on the EQ scale might be '*I can easily tell if someone else wants to enter a conversation*' and '*I really enjoy caring for other people*' (cited in Fine 2011, p. 15, her emphasis). A typical SQ question might be '*If there was a problem with the electrical wiring in my home, I'd be able to fix it myself*' and '*When I read the newspaper, I am drawn to tables of information, such as football league scores or stock market indices*' (cited in Fine 2011, p. 15, her emphasis). Baron-Cohen argues that autism is an extreme form of the male brain, and male-gendered brains produce the cognitive foundation for excellence in science and technology. He has gone on to claim that women, generally, fail at science precisely because they do not possess the type of brain that is focused on systems (Fine 2011). 'The female brain is predominantly hard-wired for empathy. The male brain is predominantly hard-wired for understanding and building systems' (Baron-Cohen, cited in Fine 2011, p. xix). According to Cordelia Fine, the author of *Delusions of Gender*, Baron-Cohen's exclusive emphasis on biological aspects of gender differences in personality and skills erases the social and cultural differences that underscore the gendering of men and women. Baron-Cohen believes that men's disinterest in empathy and interest in systems is what underscores their respective differences in attainment in science and technology (Baron-Cohen 2003b; 2000). The social and asocial personality and gender were salient features of my research at MIT.

Our starting point, then, is the skillset required to be a scientist or technologist. Sometimes the skills required to be a good or great science or technologist may not be qualities that make a person particularly well liked or socially adept. Dr. Spock, from the original *Star Trek*, or Sheldon Cooper, from *The Big Bang Theory*, are characters that find social encounters curious, difficult and infuriating, preferring instead to direct their attention to the logical perfections of science. There are some, like Baron-Cohen (1995; 2003b), who argue that asociality is, in fact, good for science and technology in that it makes better scientists and technologists. To be clear, this model of the social that psychologists like Baron-Cohen draw on is a disinterest in social interaction at the level of interpersonal communication. An example of this model was expressed in relation to George, a remarkable computer scientist known for his eccentricity.

George lived in an MIT building, and though he did not have tenure or a position, faculty members allowed him to remain in occupancy. Due to his impressive intellect, George was able to live comfortably on the prestigious monetary awards he had attained—one numbering hundreds of thousands of dollars. George had an uncertain future, in both the building and at MIT. For many years, his unorthodox habits were known but ignored by the senior authorities at MIT. But the question was asked: would his eccentric lifestyle be overlooked in the move of students, staff and faculty that was due to occur in 2004? Would his eccentric habits be overlooked in quite the same way in the new residency? George was described as a quirky individual and visionary computer scientist, promoting free software and anti-proprietary philosophies. On the other hand, George was known for having personal hygiene issues and was considered socially difficult to interact with. A member of the lab described to me how uncomfortable she felt when he spoke to her, finding his mannerisms awkward and strange.

George embodied a more extreme form of the computing personality I found at MIT and, as such, he perhaps supports the claim that there is something "different" about computer types. If there were such categories of persons as nerds or geeks, then MIT would be the place to find examples.

There is a mythical association between dirt and the rejection of norms governing personal hygiene and intellect. A famous historical example of this was the Greek philosopher Diogenes the Cynic (c.412–323 BCE), who infamously rejected Greek cosmopolitan society and lived in a tub. The mythical and cultural associations between "true" knowledge and a rejection of norms and conventions governing manners, hygiene and sociality run deep in Western cultural narratives, only to find a modern appearance in the personal and leisure modalities of nerds.

Anthropologists of science and technology dispute the absence of the "social" in scientific and technological practices drawing our attention to the presence of the social in multiple forms: sperm and eggs (Martin 1991); artificial intelligence (Forsythe 2001); nuclear technologies (Gusterson 1996); digital culture (Helmreich 1998); culture of science (Hess 1995; Law 2001) and genes (Keller 2002). For anthropologists of science and technology, the social is not exclusively the interpersonal communicative dynamics; the social is the models of race, gender, class and sexuality that permeate every aspect of existence and are then imported into scientific and technological spheres. The social is not merely reducible to the specific ways that one person interacts with another (or not as the case may be), but the multiplicity of human lived experience is sociality (Enfield & Levinson 2006).

The social that I will explore is that which is lived in the interpersonal sphere, typically the types of interactions between one person and another; but in this sphere, too, the lived existence of the human is brought to bear on interactions. The interpersonal interaction is a microcosm of multiple sociality. At the level of interpersonal dialogical interaction encounters, meetings and interactions between people are of significance and ultimately provide the bedrock of existence (Stawarska 2009; Buber 1937). The answers to these questions might rest in the types of lived social lives that feature in the day-to-day existence of research scientists at MIT. It is not unusual to see researchers using the lab like a homespace, effortlessly dissolving the boundaries between personal and public life and home and work. In the offices of NE43 (the first building residence of the CSAIL researchers), it was not unusual to see bedroom paraphernalia in offices. George, mentioned earlier, slept in his office as though it was his almost permanent residence. On the ninth floor, there was a mattress where robotic scientists would sleep, and it was not unusual to find someone napping (during the day and night) on the sofas scattered around the building. There was no clear division for many CSAIL scientists segregating home for sleeping and work for working. A typical day might start at around 10am and finish around 10pm, while for some it extended well into and through the night.

It was not unusual for some of the scientists to work through the night—though intriguingly, at the robotics lab it was the female researchers who would work through the night and even sleep there, while no men did so. One night, at 4am, I discussed this with some of them. They suggested that it might be the result of the level of relaxation the female researchers felt when working together (even under very stressful conditions). As one writer on MIT culture comments: 'The MIT student eats, sleeps, thinks, daydreams and socializes under the light of the silvery moon. The black canvas of night is the stimulus for invention' (Peterson 2003, p. iv).

Helena and Charlie would even change into their nightclothes and have toiletries and a change of clothes for the occasions when they would sleep in the lab. Helena and Charlie work continually on their projects, and Charlie in particular works constantly in the lab and claims to have worked continuously for the past 10 months. Charlie's social life is almost non-existent. It is easy to become completely absorbed in the work here. There is always something to do, demos to show (a demo is a formal demonstration of the work), code to be debugged and problems to be solved. There were occasions when I stayed in the lab for over 20 hours, sometimes for days at a time. Some of the scientists worked solidly for two days, sometimes longer. The neglect of one's personal life is sometimes reflected in the low level of vanity. Vanity is not a very useful commodity to a female scientist intent on solving the problems of her work. I did not make a value-judgment on the vanity in any gendered sense—but, when one sleeps continuously in a lab, sometimes basic hygiene standards can be compromised. Robotics is a field of technology that attracts a significant interest from female engineering and computer scientists. The robotics lab at MIT where I conducted my fieldwork had a near 50–50 ratio of males to females (though this varied according to when people's theses were completed or new researchers entered the lab). This was in stark contrast with many other groups in the CSAIL research network, which varied in the numbers of female group members with men significantly outnumbering women in the groups. When I asked the robotic professor why his lab was so different, he did not highlight it as a gender issue but instead explained the lab's gendered composition in terms of choosing the best candidates.

MIT's campus also generated a blurry sense of ontological boundaries by hosting within its perimeters a microcosm of a society. MIT's campus has many facilities so one could study, eat, exercise and socialize without ever leaving it. I want to argue here that is it not just the intensity of the work that makes computer scientists get labeled as nerds, but the very object that they engage with and the way they engage with the object when they work. Joseph Weizenbaum, creator of the computer program ELIZA, had this to say:

> . . . I know from long experience that the strong emotional ties many programmers have to their computers are often formed after only short exposures to their machines. What I had not realized is that extremely short exposures to a relatively simple computer program [ELIZA] could induce powerful delusional thinking in quite normal people. This insight led me to attach new importance to questions of the relationship between the individual and the computer.
>
> (1984, p. 7)

The community of human-computer interaction analyzes primarily exchanges between persons before programming them into scripted practices for machines (Suchman 2006; Turkle 2011; Weizenbaum 1984) and robots (Turkle 1984). These computer scientists focused on research, and though many spent long hours in the lab, not all this time was devoted to research and teaching. Some of the time was spent on socializing, doing chores, meeting with friends, eating, sleeping or exercising.

The total quantity of hours that robotic scientists would spend in the lab would span the entire day and, for some, cover the night or several days as well. Due to the quantity of hours spent in the lab, researchers rarely socialized with individuals outside MIT. When the robotic scientists were working, they were often working directly with machines—computers or robots. The researchers would spend hours writing and editing code for their robots, and often these tasks would consume their entire day.

The fact that computer scientists have unusual, perhaps "intimate", relations with machines was well expressed by video footage produced by the lab during the Independent Activities Period (IAP). The AI Olympics are part of MIT's IAP, which promotes recreational activities for staff and faculty. During the AI Olympics, computing and AI researchers are put into random groups. The researchers then script and produce a video, showing it during the final days of the IAP.

The theme of dating is unarguably the most important issue addressed in these videos. The videos are self-referential and both reflect and critique the perceptions of researchers in the labs at MIT. One film in particular featured a 'nerdy' researcher who was on a 'hot date'. The footage is ironic in several ways. Firstly, it is a female pretending to be a male, and his hot date ends up being a robot—Kismet, a former robot from the former Artificial Intelligence Lab at MIT. Kismet is an unusual creature with a gremlin-like face on a disembodied head that rests on a metal stand. As the story unfolds, the researcher is gradually instructed in how to dress, behave and perform for 'his' date. There is one scene where 'he' and a group of friends are in a bookstore, and the books they are looking at include a guide to dating and the Kama Sutra. News of 'his' progress is communicated to others via internet instant messaging. In the final stages, the 'man' has a date with the robot—they kiss, and the final credits of the films are shown with the human and robot parachuting, travelling and watching the sunset together on the beach.

GENDER AND NERDS

The relation of gender to science is well discussed in anthropology and social studies of science (Haraway 1991; Hess 1995; Martin 1991; Traweek 1988; Forsythe 2001). In anthropology, there has been a focus on the gendered nature of science, and how gendered models are infused in "objective" scientific practices (Keller 2000). The kinds of personalities that inhabit computer science and artificial intelligence became a salient issue during one series of seminars. This time, the issues of nerds and geeks related to the theme of gender and the role gender socializing played in the choice of computer sciences as a subject of study for high school students. Margaret gave a talk at the seminar series, held at NE43, that explored the issue of women and computing. Margaret was a graduate student in her twenties and her project was designing a small mobile robot. Margaret's talk outlined some of the issues, such as the statistical figures on the male-to-female ratio in computing and the efforts to encourage more women (such as through affirmative action policies).

A theme in Margaret's talk provoked a strong response from the audience members—most of whom were men. Margaret described a problem in computing, which she termed the 'nerd factor'. In some cases, women had the qualifications to apply to MIT to study computer science, but they chose not to because they felt their choice of study to be a personal sacrifice. Female high school students received unusual reactions when choosing to study computer science: 'often, the students told me, people either act surprised that they [high school women] do well in computer science or treat them as 'nerds' (Francioni 2002, p. 2).

Computer science has one of the poorest female uptakes at MIT. According to an article published in *The Tech*, an online and print magazine at MIT, female and male students in Course VI (electrical engineering and computer science) are still significantly imbalanced: 'Course VI undergraduates are, overall, 31.65 percent female. But 37 percent of VI-I (electrical engineering) students are female; 32.2 percent of VI-II (electrical engineering and computer science) students are female; and 29.1 percent of VI-III (computer science) students are female'. More astonishingly perhaps is the following figure on the attendance for higher degrees at MIT. Females make up 1885 of 4153 undergraduates (45.4 percent). When the figure is calculated at the graduate level, women only make up '1907 of 6146 graduate students (31 percent)' (McGraw-Herdeg 2008, p. 1). This has to be considered in light of the fact that MIT has roughly 50 percent more graduate students than undergraduates(McGraw-Herdeg 2008).

MIT's poor male-to-female uptake in electrical engineering and computer science is due to a number of complex factors that intersect; surprisingly, one significant factor is these subjects' perceived "image-problem"—students questioned said these subjects were for socially awkward types. MIT, Margaret explained, has tried to compensate for these issues by actively seeking female faculty and staff. The former Lab for Computer Science at MIT recommended they should 'double the number of women faculty, staff and UROPS in five years' (Howe 2003, p. 25). They were willing, but what about the other issues that prevented females from pursing studies in these fields beyond high school?

The 'nerd factor' was one important reason why women avoided studying computer science and artificial intelligence, but there were other factors that disadvantaged female applicants—private relations with machines. Margaret emphasized that the selection process for computer scientists often favored male applicants because male applicants engage in the kinds of solitary and intensive computer and robotic work that gives them the edge during the application process. Successful applicants to MIT not only score well on their SATs, but it is common for applicants to have a passion that extends well beyond their academic studies. For high school students, this can mean hours spent working at computers. Computer scientists spend many dedicated hours at computers. A professor in the lab told me a childhood story that illustrates this point well. As a child, he would constantly be tinkering with technological objects, taking them apart and putting them back together again.

There are implicit and explicit connections between intelligence and social difficulties such as autism. Are men just better at computing than women because men do not need as much social interaction? Some psychologists have tried to explain this in terms of sex differences. According to Baron-Cohen (2003a), for instance, males develop faster at 'systemizing' (S-type brains)—understanding and finding patterns in abstract systems of thought (skills that would undoubtedly give men the advantage in computer science). Women, (E-type brains), on the other hand, are more naturally empathetic. On the one hand, Baron-Cohen perpetuates naturalistic explanations for social differences and, in this respect, fails to account for the shortage of uptake of women in computer science programs. The lack of uptake on science and technology courses is not due to women's competencies in this area, but to their overall attitudes toward the discipline if they pursue the path at high school, graduate level and beyond. On the other hand, men do have some advantage over women by the time they get to university, as the source of the issue seems to be the advantage gained through time spent with machines rather than that spent in routine academic studies, and males were more likely to spend time at computers (Kubey, Lavin & Barrows 2001). Finally, some computer scientists seemed obviously uncomfortable with the discussion about the nerd factor and the image problems of computer and AI scientists. After all, this socially constructed stereotype has situated realities, sometimes with violent and, in many cases, personally distressing consequences. For many smart individuals, MIT was a safe place, where the problems of high school were forgotten. As Graham sadly reflects on his high school days:

> By singling out and persecuting a nerd, a group of kids from higher in the hierarchy create bonds between themselves. Attacking an outsider makes them all insiders. This is why the worst cases of bullying happen with groups. Ask any nerd: you get much worse treatment from a group of kids than from any individual bully, however sadistic.
>
> (2004, p. 6)

In light of the terrible experiences faced by young men and women in high school, the prospect of a campus like MIT must seem like a paradise of acceptance. If the category of the nerd is a social construction, it is a social construction with real effects in North American culture.

MIT: A DIFFERENT WORLD?

Sherry Turkle described MIT as 'a different world' when she first arrived there in the 1980s. In *The Second Self: Computers and the Human Spirit,* Turkle writes:

> Soon after I arrived at MIT, an incident occurred which captured my shock of recognition that I was in a different world. In the morning I had worked with a patient in psychotherapy who, for many months, had been using the image of 'being a machine' to express his feelings of depersonalisation, emptiness and despair. That evening, I went to a party, held to celebrate the new MIT faculty. I met a young woman, a computer-science major and one of my students, who was listening to a heated conversation about whether machines could ever think. She was growing impatient: 'I don't see what the problem is—I'm a machine and I think'.
>
> (1984, pp. 328–329)

Turkle's shock has much to do with the network she found at MIT, a network that would challenge many of her (and my) assumptions about what was human and machine, and where the boundaries and distinctions lay between them. During my conversations with AI and robotic scientists, many held mechanistic assumptions about being human, while analogies between machines and humans were constantly employed in academic and every-day conversations. For these researchers, the project to replicate a human-like robot was possible precisely because humans, too, were machines, just different kinds of machines (though arguably more complex and mysterious). The complexity and mystery of human-machines was the result of a long history of biological and natural evolution that had ultimately shaped the human-machine into the physical, mental and emotional forms it has. Human physicality, perception and proprioception results from a long evolutionary process; science and technologists, in their different ways, were finding out the biological, programmed and scripted components of what it means to be human and reshaping humanity through biological and techno-logical interventions. In turn, AI and robotic scientists were turning matter into humans (of sorts). On the one hand, robots that were as sophisticated, if not more sophisticated, than humans were a certainty; the only uncer-tainty in the process was when it would be achieved, and for this to occur, intellectual breakthroughs might be required, as well as improvements in hardware and software. Rodney Brooks sums up this perspective:

> I believe myself and my children all to be mere machines. Automatons at large in the universe. Every person I meet is also a machine—a big bag of skin full of biomolecules interacting according to describable and knowable rules. When I look at my children, I can, when I force myself, understand them in this way. I can see that they are machines interact-ing in the world.
>
> (2002, p. 174)

The mechanistic beliefs of many AI and computer scientists configure the boundaries between humans and machines in unique ways. There is no sharp dichotomous "human" and "machine", only differing perceptions, degrees of intelligence and capabilities of human and machine. Brooks continues: 'It's all mechanistic. Humans are made up of biomolecules that interact accord-ing to the laws of physics and chemistry. We like to think we're in control, but we're not. We are all, human and humanoid alike, whether made of flesh or of metal, basically just sociable machines' (cited in Henig 2007).

In social science theorizing, dichotomies between machine and human are reworked in multiple ways. As Strathern explains, 'At the same time as anthropologists have made explicit the artificial or ethnocentric nature of many of their analytical divisions, they find themselves living in a cultural world increasingly tolerant of narratives that display a mixed nature. I refer to the combinations of human and nonhuman phenomena' (1996, p. 519).

Haraway's cyborg famously debuted in her essay *A Cyborg Manifesto: Science, Technology, and Socialist-Feminism in the Late Twentieth Century* (1991). Haraway does not set humans apart from machines, but examines them as nature-culture hybrid mixtures. The famous image that accompa-nied her text in this essay is revealing. We see a woman, draped in an animal, attached to computer keyboard, with galaxies swirling in the background. Haraway's cosmology is to annihilate boundaries between humans, animals and things. Haraway is explicit in this goal. The essentialism of humanism, a belief in human distinctiveness, rational agency and radical autonomy, all marks of a superior species, is called into question. As John Gray explains in his attack on scientific humanism, scientific humanism distinguishes itself from religion, which sees humans as a by-product of a creator, and instead adheres to Darwin's theory of evolution. As Gray highlights (2002; 2011), by taking humans out of the realms of the gods and putting them in nature, they in effect diminish arguments about human uniqueness—how can humans be both part of nature and entirely different?

In this chapter, I have tried to show how there is a curious paradox of the asocial fusing with the social construction of machines. In drawing attention to the gendered differences and similarities between males and females at MIT, we have to take seriously the notion that men are viewed as asocial and women as social, yet these stereotypical notions break down at the level of "social machines". We find no such pattern when computing, AI and robotics shifts from being a predominantly male field into a field where females enter the frame. If anything, we see that trends in technology, particularly the field of "social" machines, are driven by both males and females in the field. One might even conclude that the development of "social" technologies is an outcome of more females in field. In the fields of technology, multiple, paradoxical and conflicting narratives emerge in relation to gender, sociality and asociality. These ideas are perpetuated at all levels of the lived life experience of computer science, AI and robotic researchers, and feed into the day-to-day lifeworlds of students, staff and faculty at MIT. They can be addressed via the categorization of the technological type (both male and female) as a "nerd" or "geek".

BIBLIOGRAPHY

Asimov, I 1979, *I, Robot*, Oxford University Press, Oxford.

Baron-Cohen, S 1995, *Mindblindness*, The MIT Press, Cambridge, Mass.

Baron-Cohen, S 2000, 'The cognitive neuroscience of autism: implications for the evolution of the male brain' in *The cognitive neuroscience*, 2nd edn. ed. M Gazzaniga, MIT Press, Cambridge, Mass.

Baron-Cohen, S 2003a, *The essential difference: men, women, and the extreme male brain*, Basic Books, New York

Baron-Cohen, S 2003b, 'Why so few women in math and science' in *Autism research centre papers*. Available from: <http://www.autismresearchcentre.com/docs/papers/2009_BC_WhySoFewWomenInScience.pdf>.

Brooks, R 2002, *Flesh and machines: how robots will change us*, Pantheon Books, New York.

Buber, M 1937, *I and thou*, trans. RG Smith, T. & T. Clark, Edinburgh.

Enfield, NJ & Levinson, SC 2006, 'Introduction' in *Roots of human sociality: culture, cognition and interaction*, eds. NJ Enfield and SC Levinson, Berg 3PL, New York.

Fine, C 2011, *Delusions of gender: how our minds, society, and neurosexism create difference*, W.W. Norton, New York.

Forsythe, D 2001, *Studying those who study us: an anthropologist in the world of artificial intelligence*, Stanford University Press, Stanford, California.

Francioni, JM 2002, 'A conference's impact on undergraduate female students' in *SIGCSE Bulletin Inroads*, vol. 34, no. 2. Available from: <http://cs.winona.edu/Francioni/papers/sigcseBulletin02.pdf accessed 13th November 2014>.

Graham, P 2004, *Hackers and painters: big ideas from the computer age*, O'Reilly, Cambridge.

Gray, J 2002, *Straw dogs: thoughts on humans and other animals*, Allen Lane, London.

Gray, J 2011, *The immortalization commission: science and the strange quest to cheat death*, Allen Lane, London.

Gusterson, H 1996, *Nuclear rites: a weapons laboratory at the end of the Cold War*, University of California Press, Berkeley.

Haraway, DJ 1991, *Simians, cyborgs, and women: the reinvention of nature*, Fee Association Books, London.

Helmreich, S 1998, *Silicon second nature: culturing artificial life in a digital world*, University of California Press, Berkeley.

Henig, RM 2007,'The real transformers', *The New York Times*, 29 July. http://www.nytimes.com/2007/07/29/magazine/29robots-t.html?scp=9&sq=robin%20marantz%20henig&st=cse accessed 28th December 2014.

Hess, DJ 1995, *Science and technology in a multicultural world: the cultural politics of facts and artefacts*, Columbia University Press, New York.

Houtman, G & Zeitlyn, D 1996, 'Information technology and anthropology', Anthropology Today, vol. 12, no. 3, pp. 1–3.

Howe, J 2003, 'Where are all the women?' part of "The Seminar on Dangerous Ideas" MIT. http://www.ai.mit.edu/lab/dangerous-ideas/Spring2003 accessed 28th December 2014.

Keller, EF 2000, *The century of the gene*, Harvard University Press, Cambridge, Mass.

Kelty, C 2005, 'Geeks, social imaginaries, and recursive publics', Cultural Anthropology, vol. 20, no. 2, pp. 185–214.

Kubey, RW, Lavin MJ and Barrows JR 2001, 'Internet use and collegiate academic performance decrements: early findings', *Journal of Communication*, vol. 51, no. 2, pp. 366–382.

Law, J 2001, *Networks, relations, cyborgs: on the social study of technology*, Lancaster University, Lancaster, UK.

Levy, D 2009, *Love+sex with robots: the evolution of human-robot relationships*, Duckworth, London.

Martin, E 1991, 'The egg and the sperm: how science has constructed a romance based on stereotypical male- female roles', *Signs*, vol. 16, no. 3, pp. 485–501.

McGraw-Herdeg, M 2008, 'Gender ratios vary widely across MIT courses', *The Tech* 14 October. Available from: <http://tech.mit.edu/V128/N47/women.html>. [25 July 2014].

MIT Charm School n.d., *Frequently asked questions*. Available from: <http://studentlife.mit.edu/sao/charm/faq>. [13 November 2014].

Oxford Encyclopedic English Dictionary 1991, eds. JM Hawkins & M Allen, Clarendon Press, Oxford.

Peterson, TF 2003, *Nightwork: a history of hacks and pranks at MIT*, MIT Press, Cambridge, Mass.

Picard, R 1997, *Affective computing*, MIT Press, Cambridge, Mass.

Stawarska, B 2009, *Between you and I: dialogical phenomenology*, Ohio University Press, Athens.

Strathern, M 1996, 'Potential property: intellectual property rights and property in persons', *Social Anthropology*, vol. 4, no. 1, pp. 17–32.

Suchman, L 1987, *Plans and situated actions: the problem of human-machine communication*, Cambridge University Press, Cambridge.

Suchman, L 2006, *Human and machine reconfigurations: plans and situated actions*, Cambridge University Press, Cambridge.

Traweek, S 1988, *Beamtimes and lifetimes: the world of high energy physicists*, Harvard University Press, Cambridge, Mass.

Turkle, S 1984, *The second self computers and the human spirit*, Granada, London.

Turkle, S 2011, *Alone together: why we expect more from technology and less from each other*, MIT Press, Cambridge, Mass.

Weizenbaum, J 1984, *Computer power and human reason: from judgment to calculation*, Penguin, Harmondsworth.

Films Cited

I, Robot 2004, dir. Alex Proyas.

The Day the Earth Stood Still 1951, dir. Robert Wise.

Revenge of the Nerds 1984, dir. Jeff Kanew.

Revenge of the Nerds II: Nerds in Paradise 1987, dir. Joe Roth.

Revenge of the Nerds III: The Next Generation 1992, dir. Roland Mesa.

Revenge of the Nerds IV: Nerds in Love 1994, dir. Steve Zacharias.

Television Shows Cited

Beauty and the Geek 2005–2006, The WB, USA.

The Big Bang Theory 2007–present, CBS, USA.

Star Trek: The Original Series 1966–1969, NBC, USA.

5   The Dissociated Robot

Metropolis (1927) is arguably the earliest screen classic depiction of a robot, though the term "robot" is not used in the English translation of the script. Instead, the term 'man-machine' is used to describe the robot character Maria. The creator of the robot, the wizard Rotwang, is inspired to create a race of working machines that are subservient to their human masters. In the process of creating his robot, he loses his own hand. On completing his robot, he triumphantly declares 'Isn't it worth the loss of a hand to have created the workers of the future—the machine men? Give me another 24 hours, and I'll bring you a machine that no one will be able to tell from a human being'. Cultural fictions of robots emphasize some form of loss, sacrifice or terminus. These deficiencies are reflected in these cultural narratives in multiple ways: the desire to replace an ill or disabled loved one (*AI: Artificial Intelligence* 2001) or a subservient worker population to replace human labors (*Metropolis* 1927; *I, Robot* 2004; *Wall-E* 2008).

This chapter explores how parts of self are distributed into the robotic machines. The robotic scientists at MIT developed their robots against a backdrop of personal issues involving traumatic experience, disability and loss, and these themes are prosaically reflected in the practices of roboticists, in the design of their machines and in the experiments they planned for their robots.

THE SELF IN THE MACHINE

A humanoid robot is designed to resemble a human form, but where does this resemblance begin? Perhaps the importance of embodiment in the robotics lab is such that, when robotic scientists think about the robots they will create, they think about themselves first. As the design of robots is about mimicking human forms, it seems obvious then to state that the roboticists use themselves as the first point of reference. It is not obvious to robotic scientists, though, that in crafting the humanlike form, they are drawing on their own subjectivity—their own thoughts, feelings, bodies and experiences. This subjectivity is played out in the machine in different ways that refer back to its maker—some familiar, but others unrecognizable. This is perhaps why the design of robotic machines tends to be far removed from actually looking human. Sometimes this is a conscious effort, as in the case of the labs at MIT where robotic scientists wanted to emphasize the mechanical form. But could some other effect be at play? Bakatman's remarks are useful here:

> The obsessive restaging of the refiguration of the body posits a constant redefinition of the subject through the multiple superimposition of bio-technological apparatuses. In this epoch of human obsolescence, however, a remarkably consistent imaging/imagining of both body and subject ultimately emerges.
>
> (2000, p. 98)

Bakatman suggests that, as the body becomes a site of interrogation, it is inseparable from the human subjects that inform it, which in turn are relocated on 'bio-technological apparatuses'. The robot is a bio-technological apparatus and, in this sense, biology is informed by the roboticists' subjectivities. Humanoid robots have bodies that resemble human-like bodily forms. Though robots in labs and industry can take many forms, robots of the cultural imagination are predominantly humanlike. As I commenced my research project in the field of robotics, I had expected to find full-humanoid robots in the labs at MIT. Since my first opportunity to be in the presence of a lab robot in 2003, however, I have rarely seen a complete humanoid robot that is a replica of the human body (though now many of these exist, particularly in Japan). Ironically, where they do exist, the overall humanoid form is often for "display" and often superfluous in purpose: legs that serve no function other than to give an overall appearance of the human form are a good example of this. Though you may see a full humanoid, many of the features of the robot's body add to overall form without providing function.

This MIT robotics lab is best characterized as a robot body parts lab; parts do the work of the whole (Mol 2002, Strathern 1988). The robotic scientists built body parts, not body wholes. One may conclude that, judging from the complexity of making a humanoid robot, this is a sensible strategy, and that these various parts could be reassembled to make a whole. This is not what happened in the lab, however. The robotic scientists often resisted the associative strategies to bring the different body parts together, so robot hands had no arms and robot bodies had no face or head. In only one case did the different robotic scientists come together for the development of a robot, which ultimately ended in personality conflicts and competitions of intellectual effort and time. These efforts tell us that the limits and opportunities of the robot body are intimately tied to its creators.

Regardless of the type of robot that is designed, the first point of call for its design is always the roboticist. There is intimacy in the relation between robot and human, and the researcher in creating a copy will default to the self continually through the process. If the aim is to make mechanical hands, for instance, the robotic scientist will flex their own hand and show me the features on the hand that make it important, such as the ability to grip, hold, judge weight or apply pressure. As they explain these factors to me, they maneuver and twist their own hands to show how it moves. If the robotic scientist wants to make a humanoid robot that is sociable, they think about human expression and the meaning such behavior conveys to an observer. For robotic scientists making sociable machines, facial expressions provide a plethora of communicative cues. The robotic scientists describe to me the role of facial expressions in human communication and, in demonstrating these actions to me, they stretch their lips to smile and purse their lips to frown.

When a roboticist thinks about interpersonal communication, they think of how the body is used to communicate intent and to express our interests and needs. They sometimes lift their hand and point their finger at a nearby object and wait for my vision to move to their point to show how human beings share attention with their hands, bodies and eyes. The robotic scientist's lived existence acts as the reference point for the machine. When robotic scientists think about movement, they recreate the simple sequences of behaviors involved in daily acts, such as opening a door or a jar, or expressing interest. The robotic scientists may imagine the process in their minds but, when it comes to verbally describing the events, they begin to physically act out the sequence. Moreover, in the acting of the sequence, they begin to slow down the action, pausing at each point—in slow motion, they show themselves as machines. The act of slow motion becomes an important field of capture. The technological reproduction of the senses requires the act of breaking into pieces the sound, image and text, and then speeding up the process so that these forms can be rendered in meaningful ways (Taussig 1993).

The simple sequences of human behavior become minefields of complexity for the humanoid roboticist. The language of the hand, face and body is expressed by reference to the self—yet, the language begins to change. The language acquires a technical form. In the next breath, the same discussion is repeated, but this time in the language of robotics, engineering and computer science: what mechanical systems it is necessary to employ, how many motors might the object need, what voltage is necessary to power the motors, or what circuit boards would be needed. The types of embodied practices these roboticists performed are then transformed: Using inert wires, metals, and circuits, they become mimicked in the robotic machines. As these human acts are translated into mechanical forms, the robotic scientist must then produce a machine that can echo back to them these actions.

Machined parts are prepared in advance, and come to the roboticist in fixed forms. The robotic scientists travel back and forth to the workshop to refine these devices. The metal is tethered to wires and circuit boards that allow the robot to become animate. A relation between these entities emerges between them, each part matching the other as best as possible. Robotic scientists were keen to emphasize that they were not designing robots with any special or super kinds of abilities. The objectives of their designs were to mimic and be analogous to human behaviors in the same area. Robotic scientists are faced with enormous technical issues when they begin constructing their machine, turning these ideas into a physical platform. They must have a physical robotic part of the system, and the software and computing skills to operate and control the machine. Additionally, the robotic scientist must decide and plan the behaviors of their robots. These components must work together to produce a viable operating robot.

The robotic scientist must ask: What problems will the machine be designed to address? How will the robot look? How will the robot's body be shaped? How many motors will it need so it can move? What coding systems should be used? The coding system is vital—specific languages have been designed to give the robot its behaviors. The robotic scientist must also ask how many degrees of freedom are needed for the robot to move in particular ways. Degrees of freedom (DoF) is the technical term that refers to the number of movements a robot can make and indicates the movement a body can make in a three-dimensional space. For example, if the robot can turn its head left to right, this is one degree of freedom. If it can move its head up and down this is another degree of freedom (Ross et al. 2011, p. 30). These points of movement are added up and given as an overall figure in the numbers of movements the robot can make. The more movements the robot can make, the more degrees of freedom it has. A robot entity must additionally return this movement to the spectator so that its actions can be interpreted as one specific kind of movement or another. A robot that randomly moved in any direction is disconcerting to a spectator. These types of movements must work together—a robot must be able to move its head from center to left at a required moment, otherwise the effectiveness of the movement is meaningless to human interlocutors that interact with it. The robots in this lab were designed with autonomous capabilities and, even though these capabilities were randomly arranged, they were effective in giving a sense of coherence to short encounters between robot and human. In other robot labs, bodily behaviors and facial movement were highly scripted affairs and only affected by use of remote control.

Robots require software to control their motors and actuators. The robotic scientists use circuit boards called stacks that are stacked one on top of the other, each one dealing with different kinds of information. One function might be to carry power to the motors in the mechanical body, or to instruct the mechanical part to move, or collect information from the sensors. These processes are monitored by software programs specifically designed to manage these complex systems. Programs are written in languages such as CREAL (creature) language. CREAL is a behavior-based robotics language:

> CREAL is intended for programming robots that operate for long periods completely autonomously. It is expected they will operate as

creatures in the world in that they will not be tethered (even wirelessly) to an interface, but rather will operate as they see fit.

(Brooks 2002, p. 1)

There are three main aspects to CREAL: 'a compiler that takes creature language programs and turns them into assembly code', 'an assembler that assembles code into binary form', and 'an operating system that provides runtime capabilities needed by the creature language on a particular processor' (Brooks 2002, p. 1). CREAL is a flexible computer language, as it can assemble code from different processes. In this sense, it is not a linear system but a flexible one able to compile and reassemble signals from different sources in its network. For the robotic system to work effectively, coding instructions need to be given to each aspect of the robot technology. CREAL language is incremental—that is, it has threads that can be multiplied, allowing the robot to operate using the combination of these threads. The robot's physical structure, the mechanical system, the power system, and the issues it was created to explore make up these ever-proliferating threads of complexity that must find a synthesis.

Diana E. Forsythe's *Studying Those Who Study Us: An Anthropologist in the World of Artificial Intelligence* (2001) is notable in this respect. Forsythe describes in detail the processes whereby the social aspects of AI are extradited from the final products of research papers, conference presentations and AI entities. She writes, 'The anthropologist's assertion that science is cultural is not intuitively obvious to the scientists with whom we work. For example, my informants understand of their own work as "hard science" means to them that they are part of a universal truth-seeking enterprise that is above or outside of culture' (2001, p. 2). The robotic scientists did not see themselves as "outside culture", but even though it is as important as the mechanical and electrical aspects of the science, the cultural life of robotic scientists has been neglected. I would even go so far as to say that, without cultural fictions of robots, there would be no science of robotics; the whole discipline would be constructed in entirely different ways. In some ways, the making of robots resembles the making of clock-work automata that operate on mechanical principles (Reilly 2011; Bailly 1987). The Rood of Grace (1538) was one such entity. The Rood 'had machinery that moved its eyes which caused pilgrims to see it as it if were living' (Reilly 2011, p. 20). 'The *Encyclopedia of the Nineteenth Century* offers this definition of the word "automaton"', writes Christian Bailly, author of *Automata: The Golden Age 1848–1914*, 'a machine has the form of an organized being and contains within itself a mechanism capable of creating movement and simulating life' (1987, p. 13). Automata raised questions about what was living and dead, animate and inanimate (Freud 2003), but it was the politically inspired fiction of robots by Čapek (2004) that proposed robots could be agents with the potential of political subjectivity.

As the human acts as the default mechanism and the specific lived subjectivity of the researcher, I now want to suggest that, in order to overcome this complexity, that robotic scientists pursue a specific strategy: to reflect on difference, disability and human suffering as the backdrop to making their machines. Robotic scientists import general models of disability, difference and suffering into their machines, but they also import their own unique story of suffering. Why do they choose these areas of human existence to import into their machines? The reasons are complex; in the first instance the robotic scientists import aspects of themselves into the machines they create. As the roboticists first port of call is themselves, they import their own suffering into the machines they create.

There is another more insidious reason: that in referring to disability and difference, they are making analogies between their machines and human difference and suffering such that persons with disabilities or mental health issues become intermediaries between complete humans and nonhuman robots. These models echo Euro-American notions of completeness and absence, and how bodies and selves are imagined as missing parts—a physical ability or an emotional or cognitive capacity (Ginsburg & Rapp 2013). The disabled and different are prefigured as the intermediaries between human and humanoid machines. We will begin with the first part of this strategy: the transference of psychic and physical experiential sufferings into the robots. Is this a coping strategy the robotic scientists use to deal with the complexity of the problem before them?

The robotic scientist is unconsciously engaged in a process of stamping the object with his or her characteristics. I present two case studies here. Despite all the abstract and technical processes that go into designing a robot, the result is a creature that bears an uncanny resemblance to its maker. Robots are very much like their makers.

THE BODY IN PARTS: DEFICIENT HANDS

Michael suffers from repetitive strain injury (RSI), and he is often limited in his work, as he cannot use his hands without pain. Michael is a graduate student from South America, and he had worked on many robots before settling on his robot hand as his final doctoral thesis study. During the planning stages of his project, he would use software to model all his robotic designs. On occasions, I would act as Michael's hands—following his instructions in operating the computer. Michael is making Damon—a sensual humanoid robotic hand larger than the hand of an average human male. He explained that the size of the sensors meant it had to be designed larger.

Michael's robotic hand does not match a human hand shape—it has moveable digits, rather than fixed finger positions, with flexible finger movement. Michael made a hand that can sense. Robotic sensing is fairly limited, as most robots are unable to detect any meaningful sensory data by touch. If the robot can learn to sense, it can also diversify the range of objects it can manipulate and interact with. A sensing hand would give a robot an opportunity to detect substances that are out of its range.

Michael's robotic hand is intended to be attached to the robotic arm designed by Ian. Ian designed this robotic arm for the robot Primus. This robotic arm is now used by Michael. A frame fixes the arm in position. As vision-arm-hand coordination is important, a video camera is placed on the top of the torso where a head would be. Michael is less likely to test his robots on human interlocutors. A private corporation that produces industrial robotics sponsors Michael. Michael has separated the part (hand) from the body, and he has reconnected the hand with an arm, torso and head—but these connections are merely to support the specific objectives of the hand. Like Ian, he has fashioned a body part in a vaguely human-like structure. The resemblance between the robotic body and the human body is great. Michael often had problems with his own hands, and would walk around the lab scrunching a ball to help his hand. It was perhaps a coincidence that Michael had repetitive strain injury, and he could not always use his hands to do his work. Perhaps it was a coincidence, then, that the very object he was making was a hand. Michael was not consciously inspired to make a robot hand because of his own difficulties and no connection was made. Michael conducted sensuous-vision experiments on friends and random subjects. With these subjects, he would take away their ordinary human capacities and "disable" them. Michael wanted to get an insight into how his robot perceived and sensed. By temporarily disabling his interlocutors, he was able to understand his robot's limitations and deficiencies and work out how vision and touch worked together and separately. Michael explained his work to me:

> The idea of the project is to concentrate on the use of tactile information and sensing to do with manipulation. It is different from typical research done in this area, as they rely mainly on vision for the process of locating an object and trying to do the grasping. It seems that humans do a lot more processing with tactile sensing and that is supported by many experiments. I blindfolded people and made them do things and they can actually recognize things with vision as without vision. And that's one thing. Of course we don't want to get rid of vision completely, as we want to include vision in the process; but, as a complement to the tactile and sensing . . . so we call this sensitive manipulation, because we pay a lot of attention to the sensors in the manipulation process.
>
> (Fieldwork interview 2004)

In conventional robotic design, robots are fixed to a platform and carry out repetitive actions. The definition given here is by the Robot Institute of America (Moran 2007) and shows the requirements of a robot: 'a robot is a reprogrammable, multifunctional manipulator designed to move material, parts, tools, or specialized devices through various programmed motions for the performance of a variety of tasks' (cited in Moran 2007, p. 1399). The ability of these robots to sense the environment, or to act in alternative ways, is strictly curtailed by their operational structures. As Michael explained, in order to understand his robot, he switched the deprivation to humans. Michael did experiments with human subjects that deprived them of the sensors that are ordinarily used to manipulate objects. Michael deprived them of their sight and many aspects of their touch. He describes the experiments further:

> I did these experiments with individuals and I blindfolded them and asked them to do things. . . . I put them in big gloves and made them and are only able to use two fingers. . . . So they kind of look like my robot and they do manipulation with reduced sensitivity and only with two fingers. And we did that for many reasons . . . humans . . . with reduced sensitivity they are still much better than my robot.
>
> (Fieldwork interview 2004)

Michael explains how he disables his subjects:

> The ones that I did are somewhat similar to the ones I've done in the past. The experiment consisted of the following. I blindfold them, and they sit in front of the desk. They wear a glove that's very thick and they can only use the thumb and the middle finger. They start with the right arm and they start moving. At first they start to explore the desk in front of them, trying to find objects; if they find objects, they usually try to identify them before they pick it up and then they repeat this process with different objects.
>
> (Fieldwork interview 2004)

Michael's attempts to adapt the persons in his experiments to the level of machines demonstrate an underscoring notion of a complete body that is reduced or in deficit while disabled (Ginsburg & Rapp 2013) or body parts separated from the whole of the body (Hillman & Mazio 1997; Schwartz 1997; Stevens 1997). The human is reduced in capacity to accurately represent the machine, thereby machine and the disabled body become analogous. But it was Michael's own hands that were increasingly "malfunctioning". In Michael's experiments, he deprives humans of their sensors and, in their deprivation, he creates a temporary kind of symmetry between human and robot.

THE BODY IN PARTS: DEFICIENT MEMORIES AND EMOTIONS

Helena is designing a robot with an emotional memory. She was inspired to make a robot with these capabilities because 'I used to work on a robot that was designed to socially interact with people'. In the case of Helena and her robot, she has unintentionally imbued her robot with some of her own psychological characteristics. During my first weeks in the robotics lab, Helena revealed a trauma to me that had shaped her life for the past five years. She woke one morning to find her long-term boyfriend had died next to her. He died without warning or illness from a cardiac arrhythmia, a condition of the heart. This event reshaped her life—bringing about anxiety, nightmares and depression. I asked Helena why she was so interested in memory. For her, '. . . learning and memory is very tightly integrated'. Helena explained about her work:

> My thesis project is basically development of a humanoid head robot called Radius and the idea is to have a robot that I can use to explore socially situated learning. Socially situated learning means learning from other people, or from other con-specifics in animals. So the idea is to have a robot that can be situated, socially and physically. What that means is that the robot is always there, running around in the middle of public places with people all around, so it's very important not only to have it physically, but it has to share temporal, it has to share human scale time as well. So that it's not only there for two minutes at a time every day, instead it's going to be running sixteen hours a day. It will be interacting with people all the time, and the hope is for the robot to learn from its experience whilst it's interacting with people.
>
> (Fieldwork interview 2004)

The emotional content is explained thus: 'Emotion mechanisms serve a regulatory role—biasing cognition, perception, decision-making, memory and action in useful ways' (Thomarz, Berlin & Breazeal 2005, p. 9). Helena plans to use affective tags to enable Radius to remember and distinguish between different people—taking a dislike or liking to people it interacts with regularly. Helena's project had plans that echoed her own preoccupations. Helena's therapist diagnosed her as suffering from post-traumatic stress disorder (PTSD). This event shaped her life for several years to come, and she relied on weekly therapy. Helena had spent a number of years remembering to forget, forcing the memory of the death of her boyfriend from her mind, and diminishing her emotions. The event crippled her autonomy, and there were many places she was unable to visit—the graveyard, the storage unit where she had put their belongings, and the condominium they used to share. I asked Helena about the links between her interest in memory, PTSD and Radius:

> One thing is that when I'm doing research on emotional memory, the link that keeps coming up in studies is, basically, post-traumatic stress disorder, where emotional memory is a very strong element of that particular condition. I've been diagnosed with PTSD in the past, so I don't know if that makes it a more comfortable area or maybe not so much

comfortable, but maybe it's something that I feel like I can understand more.

(Fieldwork interview 2004)

PTSD is a controversial field in the cognitive sciences. The diagnostic category emerged in the 1980s, when the American Psychiatric Association added PTSD to the third edition of its Diagnostic and Statistical Manual of Mental Disorders (DSM-III). The inclusion of this category triggered a reassessment of many conditions of previous eras, particularly the "adjustment difficulties" of Vietnam War veterans. Its inclusion was also controversial because 'psychiatrists felt that PTSD was a disparate collection of symptoms tied together only by the vague etiological feature of traumatic memory' and, therefore, more driven by political than medical science (Weizemann 1996, p. 580). While the origin of PTSD remains controversial, work into the relation between memory and emotion is considerable, and those suffering with PTSD are found to '. . . re-experience [horrific events] in nightmares, flashbacks, and intrusive thoughts . . .' (McNally 1998, p. 971).

The theme of PTSD proved to be an important theme of our conversations over the course of my fieldwork. The relation to Helena's PTSD and the systems she had created for her robot were not immediately apparent. I asked her what she thought of this relation between herself, her traumatic experience and the foundation of her work:

> Yes, I guess . . . PTSD, I guess the reason why, so there's been studies that mention increases, in some cases, memory, increases the memory process. So, if something happens to you that triggers such an emotional reaction in you, then you're more likely to remember it. But it's interesting with PTSD, the level of emotional reaction it induces in you is so great that you tend to remember it, but you don't want to remember it. That's when you start having, what do you call it, flashbacks? The affirmation of the emotional memory is so strong in traumatic experiences so the memory is there, you don't want to remember it. So it pushes and pulls in each individual. You have these flashbacks and yet you don't want to have this memory; so then, of course, once in a while you'll get these flashbacks. A flashback is basically . . . in your head, of what has happened. And this is all happening as you're trying to push it away, and the stronger you try the stronger it comes back. So that's one difference between emotional memory and PTSD.

(Fieldwork interview 2004)

Helena decided to use this model as a memory situation for Radius—forcing it to have memories based on its emotions. Radius will have to sort through a huge quantity of data that it will perceive through its visual and auditory systems. To make sense of this bombardment of sensory data, Radius will discern this material by associating positive and negative emotions with particular behaviors it perceives. The robot will have a memory and rely on affective tags to help it make sense of the overwhelming amount of information that will assault Radius's senses. Helena is also interested in error. During my time in the robotics lab, it was error that was continuously edited out of the process. Error is what a practitioner minimizes or disguises, not puts at the center of their work. After all, it is not error rates we tend to be interested in, but success rates. So why did Helena particularly concern herself with the problem of error? Why would she want to include this in her project? Already, by focusing on the question of robustness, she has moved away from her colleagues in the field—by emphasizing error, she is also reshifting the focus to the hidden and putting it on display. PTSD sufferers have repeated flashbacks and recurring memory themes. Sufferers of this condition '. . . may simply be unable to avoid recalling disturbing information; they may exhibit involuntary explicit memory' (McNally 1998, p. 1756).

Helena explained to me that her aim was to change her thought patterns. It was her thought patterns she needed to control. When her thought patterns reminded her of the trauma and the corresponding feelings, she had to reprogram her thoughts positively. Her own emotional processes were constantly subject to error—the domination of her trauma over all her other thought processes. In common with the robots she so regularly found in her studies, she was prone to error and was not robust—she could not go anywhere, she could not exist anywhere—she could only exist in specific places, like the robots she and her colleagues produced.

Helena deliberately avoided any environment that evoked a strong connection with the past, costing her considerable resources in terms of emotions and finance. Perhaps Helena's own path is echoed in that of Radius. When she was formulating ideas for Radius, a key theme became Radius's ability to exist in many different places and not be confined. Helena would need others to be with her when she went to environments she found emotionally difficult—like the assistance one gives to a robot when it cannot act alone or autonomously. Helena was not completely autonomous, either. I asked Helena what she thought of my comparison between her and her robot. I wanted to know to what extent Helena was aware of these personal influences in the making of Radius:

> That link is not so explicit in my head, but it is there. What's explicit in my head is [that] I want to work on this because I'm interested in a robot that can remember things and have emotional memories; potentially, the reason why I'm interested in it is my own experience with emotional memory.

Helena's preoccupations were refashioned as questions for her humanoid robot to address. In this case, the boundaries between Helena and Radius became blurred in the making of the robot.

DISORDERING THE BODY

If robotic design results from psychological transference from human to machine, as well as an imagination of disability and incapacity that pervades it, then what kinds of bodily forms result? What kinds of "bodies" were created in the robotic labs? The robotic scientists focus on the behavior of the body in the making of robots. These projects are all about making humanoid robots, but all are engaged in a form of deformation. The robots are incomplete. As they are constructed from metal and wires, their parts are detachable, and new parts can replace worn-out or old parts. In the case of Marius, the robot had two arms, then one arm, then two arms, then a hand on the ends of the arms, then one hand, then no hand. In the lab, the "bodies" of the robots were fragmented and made with interchangeable and transferable parts. The researchers often created the most salient parts of the body relating to the specific activity. In this context, a peculiarity occurs, and that is that the specific body parts themselves are unusual and further fragmented. Freud writes of the process of body part detachment in the uncanny:

> Severed limbs, a severed head, a hand detached from the arm . . . feet that dance by themselves . . . all of these have something highly uncanny about them, especially when they are credited, as in the last instance, with independent activity.
>
> (Freud 2003, p. 150)

The robotic scientists make hands that are larger than an average human hand or have claw-like fingers, faces without noses or mouths, eyes without faces, heads without necks, robots with torsos but only one arm, and robots with speech systems that allow them to communicate sounds but without sense. The robot is supposed to mimic a being, but these creatures look and behave very differently from an average human being. In fact, the copying of human-like characters and forms of what is human into robots parallels the production of machine objects—as objects where function and purpose are closely aligned. Researchers still claim that a meaningful exchange can take place between a human and, say, a bodiless human head, or a partial torso without legs, or a torso without a head or face-like properties. Are these robots new kinds of monsters? As Graham writes of monsters:

> One of the ways in particular in which the boundaries between humans and almost-humans have been asserted is through the discourse of 'monstrosity'. Monsters serve both to mark the fault-lines but also, subversively, to signal the fragility of boundaries. They are truly 'monstrous'—as in things shown and displayed—in their simultaneous demonstration and destabilization of the demarcations by which cultures have separated nature from artifice, human from non-human,

normal from pathological. Teratology, the study of monsters, bears witness to this enduring tradition of enquiry into the genesis and significance of the aw(e)ful prospect of human integrity transgressed.

(2002, p. 12)

In anthropology, we also find the themes of embodiment and disembodiment. Returning to Csordas (1994), he writes about how the body is a site of contestation—so, too, for embodiment-based robotic scientists. As soon as they embody the body, it becomes disembodied, fractured and disordered. Csordas writes that due to 'the destabilizing influences of social processes of commodification, fragmentation, and the semiotic barrage of images of body parts,' the human body can no longer be considered a 'bounded entity' (1994, p. 2). Haraway writes 'neither our personal bodies nor social bodies may be seen as natural, in the sense of existing outside the self-creating process called human labor' (cited in Csordas 1994, p. 2.

In the lab, a peculiar activity occurred wherein machines were carved up, even if their "body" might be extended further. The robotic scientists specifically match the robot's body to the specific act (a face is created as the center and site of a sociable robot), but also they mark this territory by making sure that no one else is able to add the same part to their robot work. The researchers acted towards the body parts in a territorial way. Marx wrote about the carving up of the body by the capitalist system:

It converts the labourer into a crippled monstrosity, by forcing his detail dexterity at the expense of a world of productive capabilities and instincts; just as in the States of La Plata they butcher a whole beast for the sake of his hide or his tallow. Not only is the detail work distributed to the different individuals, but the individual himself is made the automatic motor of a fractional operation, and the absurd fable of Menenius Agrippa, which makes man a mere fragment of his own body, becomes realized.

(1979, p. 340)

Marx recognized that the system of production under capitalism would take an individual and 'carve' up his/her body to suit the requirements of the system. In this case, it was a matter of taking something fuller and breaking it up into its smallest parts—the analogy with the slaughter house being apt. In robotics, it would be misleading to think that the scientists are working towards creating a "whole"; on the contrary, in this lab, there were conflicts over the bodies of the robots about possession and ownership. A researcher who made a humanoid torso, complete with two arms, wanted to add a robot head, but another researcher curtailed any efforts by this researcher to add a robot face to his robot body. On another occasion, one researcher reluctantly shared a robotic arm he had created. The bodily territory of the robot is divided up, dissolved and disordered.

DISSOCIATED SOCIALITY

Sociable robots also represent artificially intelligent kinds of relational arti-facts. When I describe sociable robots, I describe them as robots that have a sociable, physical and emotional repertoire that act together as a communicative apparatus. As Professor of the Robotic Life Group at MIT's Media Lab, Cynthia Breazeal's definition of a sociable robot is as follows: 'In short, a sociable robot is socially intelligent in a human like way, and interacting with it is like interacting with another person' (2002, p. 1). The advantage of this kind of reasoning about robots is that the physical embodiment of the robot agent can play a more important part in the dynamical interchange between robot and human. Imagine a human as capable of running a language program when they interact with other people, but this language program is supported, executed and enhanced by a physical body. If the person wants to communicate via textual forms, then they may require hands, arms, fingers, eyes, and a supportive framework. If a person wants to communicate via speech, they may require a mouth, a speech-box and other technical elements. But when a person is interacting via the spoken word with another human, they also execute a bodily architecture and use their bodies to send messages to the recipients (Marsh 1988; Ekman 1998). Eye contact is one essential element of this exchange (Ekman 2003).

'The eyes are the windows of the soul' is a notable English proverb, but the eyes as a significant body part occupy a place of importance in the history of the human body in culture. Artists, poets and writers have adapted this theme to communicate the importance of the eyes. 'To thee I do commend my watchful soul, Ere I let fall the windows of mine eyes' wrote William Shakespeare in *King Richard III* (1821, p. 218). The importance of the eyes is given in the New Testament: 'The eye is the light of the whole body, so that if they eye is clear, the whole of they body will be lit up; whereas if thy eye is diseased, the whole of the body will be in darkness' (Mathew 6:22–23, 1968, p. 6). Eyes provide a window into the inner workings of a person. In science, the eyes, or ocular senses, have occupied this central position in the hierarchy of the senses. The eyes are a reliable measure of objective reality, and many scientific instruments have an ocular bias, the microscope and the telescope being two of them (Edwards, Harvey & Wade 2010). New kinds of visualization techniques have accompanied these ocular sciences and facilitate the unfolding of new possibilities (Ecks 2010). Sight in this respect requires training and circumspection—seeing in science and technology is an outcome of a highly specific education in reading symbols, signs and knowledge (Cohn 2010).

For the social robot, the eyes are paramount, perhaps centrally the most important bodily part that is represented. In their hierarchy of expressions, eyes undoubtedly hold this position of privilege. Eyes give us an opportunity to fix vision, as well as a means of communicating. The use of the eyes has been important for psychologists interested in interpersonal, social and antisocial behavior. According to psychologist Peter Marsh, 'the eyes are the most reliable of all our facial features. The unselfconscious spontaneity of genuine eye signals is difficult to fake, and the smiling, darkly appraising, wistful and bored. . . . We can generally trust our readings of even subtler eye expressions. The same is not true of the mouth' (1988, p. 72).

An incapacity or difficulty in reading eyes can indicate a disability. In studies of autism, the way in which individuals do not "use" their eyes is viewed as a useful measure of someone's potential autism (Baron-Cohen 1995). In *Mindblindness* (1995), Simon Baron-Cohen suggests that eyes act as the gateway to mindreading other persons' desires and intentions (though the body can also act in this way, such as pointing with the finger or turning the body to attend to a shared object). The eyes in this sense do not only act as ocular vehicles for seeing, but also for communicating and expressing emotions. The eyes appear to communicate interior states in the absence of other physical information of the face or body being available to the perceiver (Baron-Cohen 2003).

Cultivating the eyes to see in specific ways begins from birth. Typically developing infants orient towards the face when interacting with caregivers. Visual perception in infants reveals that the action provides an important marker in the child's social and physical development with depth of focus (differentiation between different depths), visual acuity (measure of pattern recognition) and visual accommodation (capacity to focus on different stimuli in a field of vision).

In the robots Kismet, Radius and Marius, the eyes played a sensing function, detecting when people were in its field of vision. For Radius, the eyes acted as a recording device, capturing the faces of those who interacted with it. From this image, Radius applied an algorithm to sort the images, the sorting seen as a mnemonic tool to aid sociality of the creature. Face-recognition software is common in robotic toys, the recognition of the face being a key marker of the robot's intelligence and capacity to form bonds with its owners. Moreover, certain robots, such as the anthropomorphic robot NAO (designed and developed by French-based company Aldebaran Robotics) and the discontinued zoomorphic robot AIBO, a robot dog designed and manufactured by Japanese corporation Sony, can also detect other robots with their visual software.

Kismet and Radius have video cameras in their eye sockets, in the position of the iris. Robot eyes provide visual clues for attention, but their importance is to convince a person they are engaging in eye contact with the machine and are involved in a social relation. Human beings (and animals, notably primates) use the eyes as vehicles of social-communication.

What makes persons capable of reading the complex social cues of the eyes? Does it require skill and training, like the science and technologist whose material data is given in the form of codes, scripts, frequencies or waves? Or perhaps the results of an innate social architecture, built out of an evolutionary past?

In Darwin's *The Expression of the Emotions in Man and Animals* (1998) he categorizes human and animal expressions, showing the complexity of individual expressive acts that are made up of many parts, the single expression being a composite of an embodied form. Ekman claims that Darwin wrote the book to challenge a popular theory of the time by Charles Bell, who believed that 'facial muscles were given by God to express human emotions' (Ekman 1998, p. 8). Chapters are devoted to the description of the vast and complex array of human and animal expressions, such as suffering and weeping, anxiety, grief, dejection, and despair, as well as surprise, astonishment, fear and horror. The themes of emotions were taken up by Paul Ekman (2003), who categorized definite external expressions and how they correlated with internal feelings. Robot emotions are limited and scripted. Despite the limited framework of expressive emotions in robots, it is true that all manner of states are imagined in these robotic machines, even if no intended computer programming or hardware system was designed to support it. Humans have a knack at reading into robots different expressional states that may or may not be there—at least, within limits. These interactions quickly become tiresome for adults and children. Children are less patient when interacting with robots and receiving no coherent response. On one occasion at MIT, an 8-year-old boy came to visit a robot. I thought he would be excited by the prospect of seeing a "real" robot, but to my surprise his interest in the machine quickly waned when it could not perform as he expected. Adults may be more patient, then, than children in participating in the robot's development. Human emotional states are complex. Take one such expression as surprise—though it might be sequenced in the robot Kismet (eyebrows raised, eyes widened), this only forms a rudimentary indication of surprise. This is what Ekman and Friesen (1975) say about surprise:

> Surprise, for example, is an emotion with a big family. There is not one surprise facial expression, but many—questioning surprise, dumbfounded surprise, dazed surprise, slight, moderate, and extreme surprise.
>
> (1975, p. 1)

While not using the term, Darwin was describing Theory of Mind (TOM) in his general principles of human expression, which he outlines as 'the principle of serviceable associated habits' (Darwin 1998, p. 34). This state for Darwin was connected to the gratification of sensations and desires. Darwin believed that such states were also subject to the will and acts of repressions, or 'the principle of antithesis'. This state, for Darwin, is the 'strong and involuntary tendency to the performance of movements of a directly opposite nature' and, finally, 'The principle of actions due to the constitution of the nervous system, independently from the first of the will, and independently to a certain extent of habit' (Darwin 1998, p. 34). This third principle is about the spontaneous production of action due to environmental stimuli. We have, then, for Darwin the three physiological and environmental conditions that produce sets of states in humans and animals. Darwin provides a more complex description of behaviors of humans and animals, and reveals that expression is intimately tied up with the face, not as a simple device but as something complex. Take, for instance, his example of infants suffering pain:

> Infants, when suffering even slight pain, or discomfort, utter violent and prolonged screams. Whilst thus screaming their eyes are firmly closed, so that the skin round them is wrinkled, and the forehead contracted into a frown. The mouth is widely opened with the lips retracted in a peculiar manner, which causes it to assume a squarish form: the gums or teeth being more or less exposed. The breath is inhaled almost spasmodically.
>
> (1998, p. 146–147)

In this detailed description of a child crying, Darwin shows how the face, including eyes, mouth, breath, teeth, forehead, lips and gums are involved. In sociable robots, parts of the face are missing, but while the ears, nose, eyebrows, or even mouth might be missing from these creatures, the eyes or eye-like features are always present.

Ekmam and Friesen, like Darwin, analyzed thousands of photographs and drawings of facial expressions, painstakingly describing facial expressions as they appeared. For these researchers, the topic was challenging because facial expressions are difficult to illustrate, particularly as the expression can be so fleeting (e.g. a tick). Ekman saw the face as paramount in social life: it is the vehicle of communication, it is the 'site for the sense receptors of taste, smell, sight, and hearing; the intake organs for food, water, and air' (1982, p. 1). It is the central output mechanism for speech acts via the mouth and voice. The face is important in social interaction and communication and, from this perspective, it can appear reasonable for the roboticist to focus on this area at the expense of other parts of the body. The term 'Pareidolia' describes the perceptual process of seeing significant or meaningful entities in things that are not there (such as faces in clouds, or religious appearances in everyday objects) (Guthrie 1995). Why faces at all?

Faces are powerful mediators of expression. When adults engage in a social interaction, tempo and interest is mediated through eye contact, and indications of when to speak or when to stop speaking are expressed through facial directives primarily via the signals of the eyes. Embodied persons are of interest, but, as the major senses are located on the human head (sight, hearing and smell), they are exclusively mediated from these positions. Darwin believed that gestures and expressions were involuntarily used by 'man and lower animals' (Darwin 1998, p. 33) and that cross-species shared similar expressions if they shared 'analogous heads' (Darwin 1998). While Darwin attempted to show the universality of expression and emotion, he wrote these from the viewpoint of a white, upper-class gentleman of a colonial class (hence, he is less interested in men and women, but assumes men include the category of women). Darwin did try to include this cross-cultural element in his work, but relied on information from colonialists from far afield (Polhemus 1978, pp. 73–75).

In this chapter I wanted to demonstrate the interrelatedness of robot and roboticist. In designing robots, robotic scientists default to the self, and when robotic scientists are not defaulting to the self, they are defaulting to models of illness, disability and disorder. In this sense, robots are intertwined with their makers in psycho-physical terms. The bodies that are created are not whole things but fragments—built out of models of disability and mental health. The robotic scientists were often unaware that the models they used for their machines were connected with their own personal difficulties. In the case of Helena, a serious trauma affected her memory, causing her to experience malfunction, breakdown and error in her memory. These personal connections were not intentional—on the contrary, the connections between robot and roboticist are those that I observed and reflect back. In effect, symmetries emerged between human and robot, resulting in the boundaries that became permeable; yet, in the permeability it was illness, conflict and disability that were transferred to the robotic artifact.

BIBLIOGRAPHY

American Psychiatric Association 1980, *DSM-III*, Washington, D.C.

Bailly, C 1987, *Automata: the golden age 1848–1914*, Sotheby's Publications, London.

Bakatman, S 2000, Postcards from the posthuman solar system' in *Posthumanism*, ed. N Badmington, Palgrave, Basingstoke, pp. 98–111.

Baron-Cohen, S 1995, *Mindblindness*, The MIT Press, Cambridge, Cambridge, Mass.

Baron-Cohen, S 2003, *The essential difference: men, women, and the extreme male brain*, Basic Books, New York.

The Bible: a translation from the Latin vulgate in the light of the Hebrew and Greek originals 1967, Burns & Oates, Macmillan & Co, London.

Breazeal, C 2002, *Designing social robots*, The MIT Press, Cambridge, Mass.

Brooks, R 2002, *Flesh and machines: how robots will change us*, Pantheon Books, New York.

Čapek, K 2004, *R.U.R. (Rossum's universal robots)*, Penguin Classics, New York.

Cohn, S 2010, 'Picturing the brain inside, revealing the illness outside: a comparison of the different meanings attributed to brain scans by scientists and patients' in *Technologized images, technologized bodies*, eds. J Edwards, P Harvey & P Wade, Berghahn Books, New York.

Csordas, T 1994 'Introduction' in *Embodiment and experience: the existential ground of culture and self*, ed. T Csordas, Cambridge University Press, Cambridge.

Csordas, T 1999, 'The body's career in anthropology' in *Anthropological theory today*, ed. H Moore, Polity Press, London, pp. 172–205.

Darwin, C 1998, *The expression of the emotions in man and animals*, 3rd edn. HarperCollins, London.

Ecks, S 2010, 'Spectacles of reason: an ethnography of Indian gastroenterologists' in *Technologized images, technologized bodies*, eds. J Edwards, P Harvey & P Wade, Berghahn Books, New York.

Edwards, J Harvey, P & Wade, P 2010, 'Technologized images, technologized bodies' in *Technologized images, technologized bodies*, eds. J Edwards, P Harvey & P Wade, Berghahn Books, New York.

Ekman, P 1982, 'Introduction' in *Emotion in the human face,* ed. P Ekman, Cambridge University Press, Cambridge.

Ekman, P 1998, 'Introduction' in Darwin, C 1998, *The expression of the emotions in man and animals*, HarperCollins, London.

Ekman, P 2003, *Emotions revealed: understanding faces and feelings*, Weidenfeld & Nicolson, London.

Ekman, P & Friesen, WV 1975, *Unmasking the face: a guide to recognizing emotions from facial clues*, Prentice-Hall, Englewood Cliffs, New Jersey.

Forsythe, D 2001, *Studying those who study us: an anthropologist in the world of artificial intelligence*, Stanford University Press, Stanford, California.

Freud, S 2003, 'The uncanny' in *The uncanny*. Penguin Classics, London.

Ginsburg, F & Rapp, R 2013, 'Disability worlds', *Annual Review of Anthropology*, vol. 42, pp. 53–68.

Graham, E 2002, *Representations of the post/human: monsters, aliens and others in popular culture*, Manchester University Press, Manchester.

Guthrie, SE 1995, *Faces in the clouds: a new theory of religion*, Oxford University Press, New York.

Hillman, D & Mazzio, C 1997, 'Introduction: individual parts' in *The body in parts: fantasies of corporeality in early modern Europe*, eds. D Hillman & C Mazzio, Routledge, New York.

Marsh, P 1988, *Eye to eye: your relationships and how they work*, Sidgwick & Jackson, London.

Marx, K 1979, The *grundrisse*, Penguin Books in association with New Left Review, London.

McNally, RJ 1998, 'Experimental approaches to cognitive abnormality in posttraumatic stress disorder', Clinical Psychology Review, vol. 18, no. 8, pp. 971–982.

McNally, RJ 1997, 'Memory and anxiety disorders', *Philosophical Transactions: Biological Sciences*, vol. 352, no. 1362, pp. 1755–1759.

Mol, A 2002, *The body multiple*, Duke University Press, Durham.

Moran, ME 2007 Dec., 'Rossum's universal robots: not the machines'. *J Endourol*, vol. 21, no. 12, pp. 1399–402.

Polhemus, T 1978, *Social aspects of the human body: a reader of key texts*, Penguin Books, Harmondsworth.

Reilly, K 2011, *Automata and mimesis on the stage of theatre history,* Palgrave Macmillan, Basingstoke.

Ross, L, Fardo, S, Masterson, J, & Towers, R 2011, *Robotics: Theory and Industrial Application,* 2nd edn. Goodheart-Willcox Publisher, Tinley Park, Ill.

Schwarz, K 1997, 'Missing the breast' in *The body in parts: fantasies of corporeality in early modern Europe*, eds. D Hillman & C Mazzio, Routledge, New York.

Shakespeare, W 1821, 'Richard III' in *The plays and poems of William Shakespeare*, ed. Malone E. Vol X.I.X. C Baldwin Printer, London, pp. 1–299.

Shakespeare, W 1905, *King Richard III*, Blackie & Son, London.

Stevens, SM 1997, 'Sacred Heart and Secular Brain' in *The body in parts: fantasies of corporeality in early modern Europe*, eds. D Hillman & C Mazzio, Routledge, New York.

Strathern, M 1988, *The gender of the gift: problems with women and problems with society in Melanesia*, California University Press, Berkeley.

Taussig, M 1993, *Mimesis and alterity: a particular history of the senses*, Routledge, New York.

Thomarz, A, Berlin, M & Breazeal, C 2005, 'Robot science meets social science: an embodied computational model of social referencing', *Cognitive science society workshop*, pp. 7–17. Available from: <http://www.androidscience.com/proceedings2005/ThomazCogSci2005AS.pdf>.

Weizmann, F 1996, 'The harmony of illusions: inventing post-traumatic stress disorder', *ISIS*, vol. 87, no. 3, pp. 579–580.

Films Cited

AI: Artificial Intelligence 2001, dir. Stanley Kubrick & Steven Spielberg.

I, Robot 2004, dir. Alex Proyas.

Metropolis 1927, dir. Fritz Lang.

Wall-E 2008, dir. Andrew Stanton.

6   Fantasy and Robots

Blade Runner (1982) is a classic science fiction directed by Ridley Scott, and it tells the story of Rick Deckard (played by Harrison Ford). Deckard is given a mission to track and terminate four replicants, genetically engineered organic robots (you could say androids— they are more fleshy than machine). These replicants have escaped from their prescribed role and make their way to Earth to meet their creator, and Deckard has to stop them. Despite their almost identical human-like makeup, these androids are still considered different and problematic—not really human. The line between who is and who is not human and living in the film is almost absent, except that the status of these entities is different from humans. In the process of discovering who is a replicant and who is not, a test is performed on candidates: personal questions are asked and their eyes, speech and gestures are analyzed for their "human" response. In this fictional world, humanlike androids are so advanced that it takes a trained specialist to distinguish between "real" humans and "fake" ones. This does not stop Deckard from falling in love, unknowingly, with the beautiful replicant, Rachael (Sean Young). *Blade Runner* is a tale of uncertainty, boundaries and the scientific and technological mastery involved in creating artificial human beings without having the status of being human.

Robotics is a technological practice that takes place in robotic labs and is situated as part of the broader field of artificial intelligence. Robots are creatures of labs and industry, but these artifacts are also situated in fictional narratives. How, then, does the science of robotics feed into the fiction of robotics? The robot is an artifact, which is difficult to place in boundaries of any kind, and for this reason, any study of it cannot isolate one aspect of its creation. The robot is a creature of the arts, science and technology, and it can be viewed, theorized about and reinvented in a multiplicity of ways. Though I studied robots in AI research laboratories, the fictional dimensions inform and influence the context in which these objects are made, as well as how they are perceived by outsiders. I want to examine the robot in fiction as an outcome of 'transgressed boundaries' and 'potent fusions', as Haraway writes in her *Cyborg Manifesto* (1991). In this chapter, I agree with Haraway's argument that the boundary between 'fact' and 'fiction' has been breached; but where I differ is in suggesting that the boundary cannot be breached completely—it cannot be completely annihilated in the sense of done away with completely. Robotic scientists are profoundly optimistic about their practices; they describe any obstacle in robotics as a matter of the flaws in current science and technology—without questioning if their robotic projects are feasible. I ask whether it is only 'scientific barriers' that prevent AI roboticists from achieving their goals. This chapter, then, examines the robotic techniques that dispense with the boundary between fact and fiction.

Fiction is both a hindrance and a support to robotics. Cultural fantasies of robots give the work of researchers a place of importance beyond the lab, and researchers describe the robots they make as 'robots for the real world'. I heard the term the 'real world' often. On the other hand, the cultural backdrop shows the limitations of robotics work. In this chapter, I examine how the categories of fact and fiction become blurred (and affirmed) in the making of robots. There are points in the robotic science process where the practices of AI robotic science are radically different from the fictional narratives of robots. The fictional aspects were weaved into the factual aspects of the robotic science and practices.

The layout of this chapter requires some explication, as I will develop an argument that consists of multiple parts. In this sense, the chapter will read as fractured parts, yet these parts do interconnect. I begin first with an examination of the role of fantasy in the making of robots, in particular, how AI robotic scientists draw on science fictional narrative in their practices. I then shift to examine how fact and fiction has been theorized in *The Uncanny* (2003). I use *The Uncanny*, written by Freud, to consider his theories on what constitutes a boundary breach, and the consequences of such a breach. This is why *The Uncanny* is an excellent essay to examine, as it embodies several contradictions. I then explore how *The Uncanny* themes have been appropriated by robotic scientist Masahiro Mori and examine his theory of the un*canny valley* (2012), which robotic scientists use as a design strategy in the creation of robots. I spend the first half of the chapter examining the theory of the uncanny and the uncanny valley, and how these ideas are used in robotic design. Ultimately, I argue that roboticists use fiction as a way of triumphing over fear, horror and death.

The final half of the chapter is focused on the making of specific kinds of robots. The fiction does not end in the robotic practice—the practice itself is recast as a performance. In robotic circles, this is "the demo"—a demonstration of the robots' capacities. Yet robots frequently break down and do not perform their capacities—this is when the robotic scientist steps in to take their place, effectively performing as a substitute for the robotic creature. I examine the performance of robotic science practices through the robots of Primus and Marius.

ROBOTIC FANTASIES

The fact that the robot was born of a book—or, more precisely, a play—and later dispersed into many forms raises some questions about the relation of 'fact' to 'fiction'; whether, indeed, such categories exist or how these categories exist in the context of robots. Haraway argues that 'fact' and 'fiction' are interwoven, and this view is echoed by Graham, who dedicates a book to the theme, writing 'If boundaries between human, animals and machines, or between organic and technological, are clearly under pressure in the digital and biotechnological age, then the relationships between another supposed binary pair, 'fact' and 'fiction', is also central to the argument of this book' (2002, p. 13). The robot was a product of fiction, but the fiction it was born from was the lived political and economic experiences that made up daily existence for millions of men and women in the early part of the Twentieth Century (see chapter 1). Fiction does not emerge from some detached imaginary faculty of the human—fiction is a response to lived existence. This origin from fiction is important. Perhaps because the robot was created in fiction, it can never quite separate itself from fiction, regardless of what form the robot appears to take. If anything, the robotic scientists I studied moved closer to the fictional image of the robot—one with a humanoid form. Robotic fictions are intimately bound up with utopia and dystopian fantasies, and I agree with Ssorin-Chaikov when he writes of 'utopia as a practice of imagination that constitutes its own relationships' (2006, p. 357). The robot has been rendered and re-rendered in multiple ways, as an enemy, a friend, a leader, a warrior, a god, a space-explorer, a servant, a worker, a lover, a murderer, a teacher and, last but not least, a child. The robot has been replayed and readapted in fiction in numerous ways. Science and technology has an interesting relationship with fiction. In fact, it has a fiction dedicated to it—science fiction. Donna Haraway writes of science fiction as a genre that is:

> . . . generically concerned with the interpenetration of boundaries between problematic selves and unexpected others and with the exploration of possible worlds in a context structured by transnational technoscience. The emerging social subjects called 'inappropriate/d others' inhabit such worlds.
>
> (Haraway 1992, p. 300)

Around the robotics lab, it was not uncommon to see journalists, documentary makers and even film staff. Prior to my arrival, the production team and actors from the film *AI: Artificial Intelligence* had been to the lab to see the robots. Roboticist Cynthia Breazeal also acted as a consultant for the movie and still continues this connection to Hollywood productions through her links with the Stan Winston Studio. The relations between AI and popular culture are well documented. In the 1960s, scientist and author Arthur C. Clarke approached Marvin Minsky at the recently established Artificial Intelligence Lab at MIT. When Clarke asked Minsky about his predictions for intelligent machines, his vision was re-expressed in the form of HAL 9000—the disembodied supercomputer from *2001: A Space Odyssey*. In the film, the intelligence is convened by a disembodied machine—which is in keeping with Minksy's view that intelligence is independent of the body.

Even the term *robotics* was coined not by scientists, but by Isaac Asimov in the 1940s in his short story "Runaround" (1979). The parallels, then, in popular culture and AI laboratories resonate together to shape popular and scientific images of machines—so much so, in fact, that robotic scientists have to consciously alter perceptions of robots.

The following is an example of a different, but not entirely unusual, week at the robotics lab, which included a press conference. Over the course of my time here, journalists visited from several media publications including a Japanese science magazine and the *New York Times*. The press and media are a regular feature of life in robotic labs. There was an Australian documentary-maker who visited the lab in the hope of producing a biographical documentary on the life and work of Professor Kane. On one other occasion, I sat in on an interview between Professor Kane and a US production company making a film about robot science. This footage was to be part of a trailer to accompany the DVD of *I, Robot* when it was eventually released on DVD (the film was released in the summer of 2004).

The interview revealed insights into Professor Kane's theorizing about robots. The aim of the interview was to get an idea of where robots are at present, and the interviewer questioned the professor about some of the ideas raised in the film. The interviewer asked him about the relationship between fiction and science fiction and what he predicted for the future of the field. Professor Kane's response was provocative. He replied that the issues portrayed in novels and films are reflections of human desires, concerns and interests. In the future, when robots have reached a higher level of efficiency (or independence), they might not be interested in the types of issues that we think about in relation to them (the robots). Professor Kane has speculated about a 'future' where robots are uniquely different to humans, in their thoughts, behaviors and habits. He saw human theorizing about robots as limited (Personal communication 2004).

Professor Kane also mentioned during the interview that Marius the robot emerged as an idea during a showing of a film. So Marius was, in part, 'created' because of fiction. Fiction led the way and inspired science.

Robotic scientists, too, are set in the same universe as the outsiders to the lab, and they do not relate to the science of robotics in technical terms exclusively. The researchers also display features of the uncanny as they describe and find meaning in the work they produce. Prior to their work as robotic scientists, they had related to robots as cultural objects before technological ones. Is a kind of Utopianism at work? Utopian conceptions generate 'a sense of hope for the future, thereby produce the essential preconditions for human actions which aim to transform the world in accordance with an image of what it should be' (Meisner 1982, p. 4). To illustrate these themes, I want to draw on the autobiographical themes that act as the backdrop to the making of robots.

It is important to note here that robotics is an international practice. In one robotics lab, the members were drawn from eight countries. The connection with Japan is even more important for these researchers, because Japan is at the forefront of making humanoid robots. The traffic between Japanese corporations and universities flowed in both directions—lab members would visit robotic labs in Tokyo, while visiting researchers and professors would work at MIT. If anything, the robotic researchers were envious of the Japanese developments in robotics—and equally envious of the cultural acceptance of robots, a factor the non-Japanese researchers felt was absent from their cultures. In the following, I describe symmetries between childhood and adolescent fantasies and the making of robots. This is demonstrated by interviews with leading robotic scientists. It is the roboticist who is continually drawing on fiction as the backdrop to the making of robotics. I asked a famous Japanese roboticist to tell me his motivations:

> In the beginning, I have to say something about my history. When I was a small kid, maybe ten years of age and after the end of World War II. Japan, after the war . . . was totally destroyed. However, many people struggled to recover in Japan. And years later, economic prosperity started. So, for example . . . TV programs began broadcasting . . . about ten years later at the end of World War II. There were a lot of trial TV programs, dramas . . . comedies and one of the interesting things is animation. At that time only animation was played in movie theaters. I believe that the first animation was *Astroboy*, especially made before TV broadcasting. So I watched the TV and also there are a lot of other types of robot animations all over Japan. So that was so exciting. I was totally impressed by that kind of movie . . . cartoons . . . animations. And also, at that time there were also a lot of science inventions and technological prosperity and advancement. At that time . . . the US competed with the Soviet Union for . . . space exploration, such as the launch of the rocket to the moon. So, Japan was excited about scientific advancement. So was I. Do you know Manga? I read a lot of Manga when I was a kid. On the first ten or twenty pages there are a lot of pictures about our future. So, thirty years later, cars should fly and not moving along . . . many people have big hopes for the future of science. So many boys have a lot of dreams about the future and are interested in technologies. I'm one of those boys. It is an interesting history for me, usually if you're a graduate in primary school, you usually write some essay 'my dream in the future' or something like this. So I wrote

a simple paragraph: "I will become a robotic researcher [and] go to university.

(Personal communication 2003)

There are a number of fascinating themes raised by the professor of robotics. The relation of his own childhood fantasy of a 'future' society with robots and flying cars was realized in his activities as an engineer. Moreover, his childhood fantasies acted as the backdrop to his adult fantasies and informed the direction of his work. The professor did not begin making robots, but his research led him to make robots, and he describes what he does in terms of it as connected with his personal narrative of self. There is a sense in his story that his technological activity was an unfolding of his inner desires—a sense that his own desire and fantasy would be realized in the form of robots. The biographical narrative is structured like a fiction—with a hero realizing a dream. The Japanese robotic scientists' account might be due to what David Bleich (1984) describes in relation to fantasy:

> The principle is that a motivating fantasy can be discovered as the common ground between the personal lives of cultural leaders, works of imagination read by the public in general, and more abstract intellectual communities of the age. A fantasy is, perhaps, an operational definition of a feeling. It is a way of naming the feeling in dynamic or behavioral terms rather than in simple denotational terms like 'love' or 'fear' or 'anxiety.'

(p. 2)

In my own work, there were interlocking relations between 'fact' and 'fiction', and claims about AI often took on a fantastic quality as though these technological narratives had inherited the architecture of fiction—where fantastic possibilities exist outside the realm of natural, social and personal relations.

THE UNCANNY

Freud's 1919 publication *The Uncanny* is a study of the breakdown of the boundary between fact and fiction and how they interplay. Freud expanded on the theme of the uncanny first raised by Ernst Jentsch in an earlier essay in 1906 (Jentsch 1997). Freud (2003) describes the uncanny in a variety of ways: it is 'the unhomely' (p. 152); the difficulty in judging 'whether something is animate or inanimate' (p. 147); the confusion when something 'bears an excessive likeness to the living' (p. 147); that which provokes 'intellectual uncertainty' (p. 146) or an 'inner compulsion to repeat' (p. 145); and it is also a consequence of 'the double' (p. 141). Freud was interested in the context that would facilitate this breakdown: how it would occur and how it might be explained. Before examining what triggers the uncanny, I want to first examine the various conditions that do not trigger the uncanny:

> Indeed, the fairy tale is quite openly committed to the animistic view that thoughts and wishes are all-powerful, but I cannot cite one genuine fairy tale in which anything uncanny occurs. We are told that it is highly uncanny when inanimate objects—pictures or dolls—come to life, but in Hans Andersen's stories, the household utensils, the furniture and tin soldier are alive, and perhaps nothing is further removed from the uncanny. Even when Pygmalion's beautiful statue comes to life, this is hardly felt to be uncanny.
>
> (2003, p. 153)

The uncanny, then, is not triggered by the activity of animation of the inanimate per se, but the animation of the inanimate in a context. 'This suggests that we should distinguish between the uncanny one knows from experience and the uncanny one only fancies or reads about', writes Freud (154). When does the boundary, as Freud describes in *The Uncanny*, get so blurred that the uncanny is triggered? Freud included psychical and physical states in *The Uncanny*, and referred to numerous conditions that might trigger these states: the fictional, imaginary, physical and real. Robots had not been invented by the time *The Uncanny* had been published, but Freud examines the issues raised by human-like objects, such as automata and dolls. The essay then serves as a device to analyze psychical states and artifacts that destabilize categories.

Automata were objects that some thought destabilized the boundaries between human and machine, living and dead, animate and inanimate (Reilly 2011; Bailly 1987). In the seventeenth and eighteenth centuries in Europe, hundreds of mechanics constructed human and animal automata and animal automations, such as Jacques de Vaucanson's Duck, which was exhibited in the 1700s and was said to drink and defecate. The automata gave rise to the term androïdé, defined as 'an automation in human form, which by means of well-positioned springs, etc. performs certain functions which externally resemble that of man', as it was cited in Diderot and d'Alembert's *Encyclopédi* (see Standage 2002, p. 20). Čapek's robot later replaced the androïdé in popular usage to describe a human-machine. The making of automations raised questions about the boundaries between human and machine. The ability to do 'technologically' what 'nature' has only done in the past creates fears that anthropologist Mazlish describes here:

> . . . the fears of the automata . . . posed an 'irrational' threat to humans calling into question their identity, sexuality (the basis of creation?), and power of dominations. Hence the man-machine is a paradox that

embodies both the potential for the creative Promethean force as well as fear and trembling.

(1995)

The automation known as the Turk added to these themes. The Turk was designed by Wolfgang Von Kempelen and exhibited through Europe and the US in the 1700s. The Turk was a distinctive automaton because it played chess, fooling audiences into thinking it was a 'living' machine as the Turk was perceived to 'think'. 'By choosing to make his machine a chess player, a contraption apparently capable of reason, Kempelen sparked a vigorous debate about the extent to which machines could emulate or replicate human faculties', writes Standage (2002, p. xiv). The Turk was shown to European and American audiences, and it was so popular that one newspaper wrote 'This wonderful piece of mechanism continues to attract full houses by day and night' (Standage 2002, p. 151). The Turk was later revealed to be a fraud, and the claim of a chess-playing automaton proved false. The Turk's chess-playing abilities came from a man positioned inside, controlling the actions of the automata (see Standage 2002). The Turk, at least conceptually, fulfilled the conditions of the Turing Test, even though it lacked any AI. The Turing Test claimed that machines were intelligent if humans failed to discern a difference between human and machine responses (see Malik 2000; Standage 2002). The making of automata raised questions about what was human or machine, living or dead, animate or inanimate. Automata produced 'uncanny' effects in their audiences. Freud drew on the history of automata to explore the relations between persons and things in *The Uncanny* which, in one sense, was 'that species of the frightening that goes back to what was once well known and had long been familiar' (2003, p. 124). Many automata were anthropomorphic mimetic objects that expressed this 'something' that was both familiar and unfamiliar. As Gaby Wood explains '. . . there was anxiety in the present situation—an anxiety that all androids, from the earliest moving doll to the most sophisticated robots, conjure up . . . a perfect reaction example of what Sigmund Freud called "The Uncanny", the feeling that arises when there is an "intellectual uncertainty" about the borderline between the lifeless and the living' (2003, p. xiv). Freud describes the uncanny as an 'eerie-feeling' when one has 'intellectual uncertainty'. Freud explores how fiction is able to confuse boundaries and purposefully conjure uncanny relations:

The uncanny that we find in fiction—in creative writing, imaginative literature—actually deserves to be considered separately. It is above all much richer than we know from experience; it embraces the whole of this and something else besides, something that is wanting in real life. The distinction between what is repressed and what is surmounted cannot be transferred to the uncanny in literature without substantial modification, because the realm of the imagination depends for its validity on its contents being exempt from the reality test. The apparently paradoxical upshot of this is that many things that would be uncanny if they occurred in real life are not uncanny in literature, and that in literature there are many opportunities to achieve uncanny effects that are absent in real life.

(2003 pp. 155–156)

In the state of the uncanny, the person loses their capacity to judge what is fact or fiction, living or dead, animate or inanimate. The categories of human or non-human, living or non-living lose their comfortable placement in the realm of the uncanny (Wood 2003; Mori 1999). Freud described the uncanny as terrifying:

There is no doubt that this belongs to the realm of the frightening, of what evokes fear and dread. It is equally beyond doubt that the word is not always used in a clearly definable sense, and so it commonly merges with what arouses fear in general.

(2003 p. 123)

Humanoid robots are objects that traverse and renegotiate what is human and non-human, and I am interested to know what Freud might say if the boundary between fact and fiction was itself reworked, as it is through social anthropological theorizing. How might the contemporary reworking of the boundaries between fact and fiction operate in the context of the uncanny? Will it err towards the fairy tale or the horror story?

ROBOT DESIGN: THE TRIUMPH OVER DEATH

Masahiro Mori is perhaps the most important contributor to issues of humanoid robotic design, writing in the early 1970s, when robots predominantly were made by quirky artists and amateur technologists (Reichardt 1978). In *The Buddha and the Robot*, Mori proposes an approach to these objects that is furnished by spiritual meanings, showing how technological ideas can be compatible with Buddhism. However, Mori is more known for his ideas of the uncanny valley. Mori appropriated the themes of Freud's uncanny and readapted these themes to the making of humanoid robots. The uncanny valley was a theory of robotic design developed by Mori in the 1970s. Robotic scientists would frequently reflect on the uncanny valley. The uncanny valley puts human-like objects on a grid, with appearance at one axis and behavior at the other. Mori scaled objects up and down the grid, comparing the behavior and appearance of artifacts; for example, a zombie would be placed in the uncanny valley, because a zombie is ambivalent, dead and moving simultaneously, engaged in acts that are diametrically opposed categorically. Mori reasons that, if certain objects provoke discomfort, even in their imaginary form, then being conscious of design could help humans accept robots more easily. Mori's argument, then, is that the more intelligent robots become (their behaviors) the more their appearance must remain different—putting the metallic and machine nature of the robot on view. Alternatively, if robots look too human-like but they do not behave 'intelligently', then this can also create problems, as the robots are perceived to have more intelligence than they might possess. Helena, a robotic scientist, told me about Mori's theory of the uncanny valley:

> I would not call it a theory. This is something that a lot of people have criticized. Mori never actually carried out any experiments to support his "uncanny valley" concept. It's merely a hypothesis. He's never shown any results or data to suggest that it is in fact true. I agree that many roboticists have conceded to the idea but certainly do not believe it as 'fact'.
>
> (Fieldwork interview 2003)

Mori's ideas are pervasive but not unquestioned in the field of robotics. A research project at the University of Texas at Austin run by David Hanson aims to challenge Mori's view of robotic design by designing robots that are "realistic". Unlike Hanson, the researchers at MIT deliberately designed their machines with their mechanical nature on show, as the robotic scientists saw problems with designing machines that looked too human-like, such as the effect that might be created with using a skin-like fabric on the surface of the machine, like that used by roboticist David Hanson. Mori's motivations for the uncanny valley were different from Freud. Freud explored the peculiar state provoked by a conflation of boundaries between fact and fiction— the kind of fears, terrors and fantasies provoked by such phenomena. Mori, on the other hand, was a roboticist, and his theory was a design philosophy aimed at first understanding the conditions that provoke the uncanny, and then pursuing techniques that would overcome these issues in the context of robotics. It is interesting to note how Mori might think that a design strategy might compensate or minimize the impulses and psychical and emotional states that Freud writes about in *The Uncanny*—which has been the subject of countless texts thereafter. More unusual is perhaps the appropriation of a psychoanalytical theory to robotics—which is, in effect, what the uncanny valley aims to do. Mori's own descriptions of the uncanny follow a similar pattern to Freud's, where he indicates the kinds of events, states, processes or objects that might provoke the uncanny valley, particularly the theme of the dead. It is interesting that he chooses the dead to illustrate his point—placing the dead in the same valley as certain kinds of robots. Perhaps this is Mori's own way of overcoming death—if not through a change in human circumstances, then in terms of a robot's creation—designing it in specific ways so that it does not fall into the valley with the dead:

> A healthy person is at the top of the second peak. And when we die, we fall into the trough of the uncanny valley. Our body becomes cold, our color changes, and movement ceases. Therefore, our impression of death

can be explained by the movement from the second peak to the uncanny valley as shown by the dashed line in the figure. We might be happy this line is into the still valley of a corpse and not that of the living dead! I think this explains the mystery of the uncanny valley: Why do we humans have such a feeling of strangeness? Is this necessary? I have not yet considered it deeply, but it may be important to our self-preservation.

(2012, p. 35)

The robotic scientists I met at MIT knew of and used Mori's ideas in the design of their robots—they designed robots that took on the outline of a human form but kept the features of their machine-ness explicitly on display. The intention of these robotic scientists is to minimize public concern and many robots that use Mori's assumptions also are robots that have had ongoing public exposure. In using Mori's design strategies, the robotic scientists hope to minimize feelings of discomfort and concern that might be triggered when seeing or interacting with humanoid robots. Robotic scientists may try to design their robots with a conscious intention to minimize discomfort, fear or intellectual uncertainty; yet, the robots they create rarely achieve this regardless of how the robot is designed, whether with metallic or skin-like surfaces. It appears that, in spite of what robotic scientists try to do to help human interlocutors feel comfortable with machines, the uncanny is an unavoidable consequence of this interaction. Why is this? The robotic scientists I studied employed Mori's design guide in the hope of making robotic artifacts sit comfortably with humans. Yet, the reactions of laypersons seem to suggest that the uncanny is something that is provoked by artifacts that destabilize boundaries; if this is so, there may be little robotic scientists can do to help humans feel more comfortable with their artifacts. This does not stop Mori from suggesting that the categories of living and dead are ambiguous and contingent:

A dead person's face may indeed be uncanny: it loses color and animation with no blinking. However, according to my experience, sometimes it gives us a more comfortable impression than the one given by a living person's face. Dead persons are free from the troubles of life, and I think this is the reason why their faces look so calm and peaceful. In our mind there is always an antinomic conflict that if you take one thing you will lose the other. Such a conflict appears on one's face as troubles, and makes his, or her, expression less comfortable. When a person dies he, or she, is released from this antinomy, and has a quiet expression. If so, then, where should we position this on the curve of the uncanny valley? This is an issue of my current interest.

(Mori 2005)

Mori puts 'living human beings on the highest point of the curve', but then goes on to reflect that human beings may not be alone in this position.

Besides his attempt to overcome death through the critical design of robots, he also speculates about the ultimate transcendence—that of gods:

> It is the face of a Buddhist statue as the artistic expression of the human ideal. You will find such a face, for example, in Miroku Bosatsu (Maitreya Bodhisattva) in Kohryuji in Kyoto, or in Miroku Bosatsu in Chuguji and in Gakkoh Bosatsu (Candraprabha) in Yakushiji in Nara. Those faces are full of elegance, beyond worries of life, and have aura of dignity. I think those are the very things that should be positioned on the highest point of the curve.
>
> (Mori 2005)

Mori's view that the more a robot exhibits its own mechanical nature, the more it is a guard against the uncanny, seems to have little impact in real settings. The use of Freud's theories in robotics reveals their complexity. Is it possible that these design strategies really compensate for the uncomfortable fear that Euro-American persons seem to experience in the light of boundary-blurring processes, states and objects? First time viewers still described their responses in a way that resembled the kinds of issues that Freud raised. Here is Gaby Wood's impression of Cog at MIT: '. . . when they are working there is a creeping sense of a living presence. Cog, a monumental metal torso on a three-foot pedestal, follows me around the room with its eyes . . .' (2003, p. xx). Cog, short for "cognition", was a humanoid robot at MIT's former AI Lab. Cog had limited facial expressiveness, yet just the substitute "eyes" of the cameras could be enough to make audiences think it had internal states, and laypersons often felt the robot was looking at them:

> Cog 'noticed' me soon after I entered its room. Its head turned to follow me and I was embarrassed to note that this made me happy. I found myself competing with another for its attention. . . . I had heard Rodney Brooks speak about his robotic 'creatures,' I had always been careful to put quotation marks around the word. Which now, with Cog, I had found the quotation marks had disappeared. Despite myself and despite my continuing scepticism about this research project, I had behaved as though in the presence of another being.
>
> (Turkle cited in Brooks 2002, p. 149)

When seeing the robot Marius in action I, too, had an uneasy feeling. The robot had animate movements and it moved its arms and head. I was impressed by its movements and tried to interact with it. I found myself alone in the room with it and at times felt uncomfortable, yet at other times hardly noticed the robot. It was sometimes just as disturbing when inanimate as it was when animate. When it was moving, its behaviors appeared random and, after the initial "wow" factor, the behavior appeared to have little purpose. I remember spending some time "talking" to Marius and showing it objects, usually brightly-colored toys that littered the room. Marius was in a room that acted as a corridor, and so researchers and staff had to pass by the robot to get from one side of the lab to another. The room in which Marius was based also had a sofa, and I would look at Marius as I worked on my fieldnotes. Its inanimate stature was sometimes just as disturbing, as it sort of looked human but was totally still. I found my mind wondering what would happen if it suddenly came alive of its own volition—which it never did. It was still and eerie. When it moved it was eerie. I could see it was a robot—its mechanical nature was on the surface—yet the uncanny was provoked. At other times, though, I was bored by the robot. It just stood there and did nothing and, though it may have had dozens of wires and cables and sophisticated machinery surrounding it, it was often just "dead"—turned off.

The robotic scientists had designed Marius according to the design principles of Mori, or at least paid lip-service to the concepts of Mori's uncanny valley. Mori's design strategy is no guarantor in preventing the uncanny— and the robotic scientists may not be able to transcend death, nor the state of the gods, by their technological designs.

PRIMUS IN THE REAL-WORLD

In the robotics lab, I never failed to hear about fiction or reality—categories that were repeatedly expressed. Despite all the fantasy, fictional and popular cultural influences that circulate in the robotic scientists' socio-personal context, there is also another "world" that circulates—that of the 'real'. In robotics, researchers often speak about the real-world—the robots have to work in the real-world. This term is so important for describing the eventual places robots will occupy that the term real-world is used capriciously, and any demonstration or explanation of the robot is usually accompanied by its importance in this real-world. The robotic scientists speak continually of designing 'real-world' robots, or 'robots for the real world'. What is this real world? What would these real world robots look like or do? What is the function of the real in the making of robots?

In the field of behavior-based AI, there is an argument about how to get the "world-out-there" into machines. In the context of the robotics lab, real-world robotics referred to several different things: it implied a robot would be autonomous, it implied a robot would respond or behave in a way that a human might, and it implied that the robot would operate in the human-designed built environment. Getting a robot to act in the real-world is closely connected with the technologies of subsumption architecture designed by Brooks (1991). Below is a description of how subsumption architecture helps robots operate in the 'real world'.

Subsumption architecture is designed to deal with a limited part of the real world, in particular robot navigation and obstacle avoidance. The architecture uses the real world as its own model, thus avoiding problems of maintaining an internal model of the world. Problems in dealing with the dynamics of the real world are avoided by not attempting to predict anything about it. Subsumption architecture expects imperfect data from its sensors. For this reason, the real-world acted as both a guide and restraint to the work in the lab. I want to examine how the real-world guided and restrained the activities of the robotic scientists by examining the making and abandonment of the robot Primus.

In order to understand this approach in the context of robotics, it is worth recapitulating some of the issues raised in chapter three regarding the contest between traditional and embodied robotics. In traditional AI, the sentiment is that reason can be captured by rules, and as more sophisticated algorithms are developed, then such algorithms can accommodate and overcome changes in the physical environment. Embodied behavior-based robotics, however, is formulated and developed on the assumption that the environment is unpredictable and, by giving robots bodies with sensors and autonomous capabilities—as in subsumption architecture (the practice where modular components work together to create larger systems)—these robots can have a greater impact in the world with fewer programmed capabilities. Rodney Brooks' approach to solving these problems involves creating robots with sensors:

> In these new approaches there is a much stronger feeling that the robots must find everything out about their particular world by themselves. This is not to say that a priori knowledge cannot be incorporated into a robot, but that it must be non-specific to the particular location that the robot will be tested in. . . . this forces behaviour-based robots to operate in a much more uncertain, and much more coarsely described world than traditional AI systems operating in simulated, imagined worlds.
>
> (2002, p. 75)

This means that rather than inscribing the "world-out-there" into machines, machines could interact with the real-world if they had the physical equipment to interact with the world through sensors, and behavior-based embodied structures. The robotic scientists continually insist that they are designing robots that will need to interact in the real-world. Therefore, their measure of their success is an autonomous robot that can meet its goals without any human assistance.

The lab partook in a national project sponsored by the Defense Advanced Research Projects Agency (DARPA), who provided researchers with a mobile robotic platform. The platform was called the Segway Robotic Mobility Platform (RMP). The platform was developed by Segway, LLC with DARPA funding. In collaboration, Segway, LLC and DARPA manufactured specifically modified versions of Segway's Human Transporter—and it was this modified platform that lab members used as the basis to build a robot. In discussions preceding my fieldwork, the group had an opportunity to partake in a national project to build a robot on a mobile platform. Simultaneously, and initially independently of this opportunity, two robotic scientists began to think about different kinds of robots to build in order to explore different issues in humanoid robotics. Graduate roboticist Helena wanted to build a socially-interactive humanoid head. The professor of the lab was interested in mobility, embodiment and behavior-based robotics. The opportunity, then, to build a robot that had mobile capabilities provided an opportunity to explore the elements of interest amongst the researchers. Luke, a post-doctoral roboticist, contemplated building a three-armed robot when reading about one in a science fiction story. Luke is an avid science fiction fan and drew his inspiration from a book that was about a race of three-armed aliens called 'Moties'. As Luke explains:

> Niven and Pournelle in *The Mote in God's Eye* write about a race of three-armed aliens called the 'moties'. They have sub-races that are excellent engineers, warriors, and diplomats. That's where I drew that name from.
>
> (email communication, 3 July 2003)

There were several informal discussions about what the robot should be called. There were three names in competition for the new robot—Vishnu, Dexter and Motie. Motie was suggested by Luke. In the end, the robot's name was neither Vishu, Dexter nor Motie, but a classical Roman name. In my thesis, I call this robot Primus to protect issues of confidentiality.

The robotic scientists speculated about what a three-armed robot could do. If it had three-arms, it could hold an object in place with two arms and use its third to manipulate the object. If it had a sociable head, it could interact in sociable ways with people. If it had a mobile base, it could move around autonomously by sensing the world. If it had all of these things, it would be human-like in simple ways. The robot's progress became centralized during regular group meetings.

Primus's goal was to show how an autonomous mobile robot could open doors. The group wanted Primus to autonomously navigate along corridors, detect doors, and to open and go through doors without relying on remote control or pre-programming. Primus had a sonar system to prevent it from bumping into walls and doors, and several sonar detectors were placed on the mobile robotic platform.

Primus also had a vision system to help it navigate along corridors and differentiate a door from a wall. Thomas, an AI graduate student in his early thirties, explained the problems of vision to me. Thomas described how vision systems are biased to the time of their creation. If, for example, the system was designed during the mid-day sun but the machine then had to work in the dead of night, the machine's 'world-out-there' would have undergone changes. In the field of machine vision, designing programs that would help a machine system distinguish between a wall and door is profoundly difficult. In human terms, distinguishing a door and wall is straightforward unless affected by impairment. In human visual systems, doors and walls are different types of things—though they may share the same color or texture. The human visual system's ability to detect texture, color and depth combine together. In machine vision, how does one include a category like texture in the visual program? The problems for the machine vision experts is further increased by the subtle variations of color that add shading, shadows or alternative stretches of color to the same space.

THE THEATER OF THE LAB AND
THE ROBOTIC PERFORMERS

In the case of Primus, its performance goals were rarely achieved. The corridor was only a few meters long, yet the robot could rarely navigate along the corridor, detect a door, or use an arm to open and go through the door. The robot was placed on a wheeled base, which allowed it to navigate in spaces. When Primus failed to perform its actions, the robotic scientists took the place of the malfunctioning artifact, and so the robot's perception was demonstrated through the human example. In the field of AI robotics, I came to realize the importance of the demo. The demo is a demonstration of the robot's abilities. The roboticist will aim to demonstrate the robot's abilities, but it can substitute video footage of the robot's performance in place of an actual performance. Therefore, the demo has come to occupy an important place in robotics. Helena explained:

> In robotic research, typically what happens is you get the robot to do a particular task, to manipulate or to walk, or to interact with people. And when you do that, you show the results of the robot doing that for a minute or less, or slightly more when you tape that on video and show the results. It's not uncommon that these results are not reproducible, for example if you need the robot six feet to the right some results might not be reproducible [or] if you change the objects that the robot use, [or] if you use different people to interact with the robot, it might not be able to produce the same results.
>
> (Fieldwork Interview 2004)

During demonstrations—or demos—the roboticist programs his or her machine to act out its activities. In some cases, the robot malfunctions and the systems fail to work as intended. When this occurs, I have seen the roboticist act out the activity instead. The making of humanoid robots is a practice in translation; it is directly concerned with refashioning the human in mechanical forms. In the context of robotics, the object's lack of capacities merely allowed a human stand-in to act out the performance in its place. In this way, the roboticist became a stand-in for their robot. When video footage of the robot's abilities is taken, researchers only emphasize the footage where the robot achieved its intended goals. The 'real-world' is ultimately inseparably infused with fictional textures.

Robotic scientists have to regularly show their work to outsiders. When the roboticist begins his or her talk, they describe in advance what we can expect to see the robot do and why the area of research was chosen. The roboticists use their own bodies to act out the human performance they want to have their robots perform. These actions are replayed over and over again to different audiences—representatives of funding bodies, members of the public, or the media, and to friends.

A group member will lay in wait with a video camera, ready to capture its performance. As the making of a robot is an engineering-based, experimental activity, it is not surprising that the robot should have to undergo many trials and reprogramming before it can do what was intended. I draw attention to its "failure" not as a criticism of the practice, but as a criticism of the claims made of the robots in excess of the practice. Researchers excused the failing of Primus in their own way. The robotic scientists believed that, if they had worked as individual researchers (and therefore recouped the prestige of individual work), the project would have been more of a success. The group had several conflicts over the robot during this period, leading one researcher to leave the group. The researchers were hostile to group work and preferred to work independently. As Primus required the efforts of many group members, it was not treated as important.

The researchers eventually got Primus to navigate along a corridor, open a door and go through it. As the pressure on the researchers to make the robot perform increased, the researchers relied increasingly on a rule-based AI. The team began to measure and reset the platform with each attempt to make it perform its action. Primus did achieve its goal, but it did this with substantial modifications to the underlying theory and, rather than the real-world aiding the project, it acted as its constant barrier to success.

The robotic scientists began to compensate for the failings of the robots by acting out what the robot should do, not only for Primus, but for all the robots. It was the robotic scientists who began to fill in the gaps of the robot's failings. In this case, the robotic scientists turned on each other rather than the theory. No one questioned the underlying theory—that remained intact as group members blamed each other for the failure of the robot and the project in general.

As the DARPA deadline approached, there existed a gulf between Primus imagined and Primus created. As the weeks went on, the ambitions of the robotic scientists were gradually winnowed down from a three-armed robot to a two-armed robot, and then finally to a one-armed robot. After its one-time only performance for the funding agency, Primus was put aside, referred to occasionally as an unfinished project.

BIBLIOGRAPHY

Asimov, I 1979, *I, Robot*, Oxford University Press, Oxford.

Bailly, C 1987, *Automata: the golden age 1848–1914*, Sotheby's Publications, London.

Bleich, D 1984, *Utopia: the psychology of a cultural fantasy*, Ann Arbor, Michigan.

Brooks, R 1991, 'How to build complete creatures rather than isolated cognitive simulators' in *Architectures for intelligence*, ed. K VanLehn, Lawrence Erlbaum Associates, Hillsdale, NJ, pp. 225–239.

Brooks, R 2002, *Flesh and machines: how robots will change us*, Pantheon Books, New York.

Freud, S 2003, 'The Uncanny' in *The Uncanny*, Penguin Classics, London.

Graham, E 2002, *Representations of the post/human: monsters, aliens and others in popular culture*, Manchester University Press, Manchester.

Haraway, DJ 1992, *Primate visions: gender, race, and nature in the world of modern science*, Verso, London.

Jentsch, E 1997, 'On the psychology of the uncanny (1906)', *Angelaki: Journal of the Theoretical Humanities*, vol. 2, no. 1, pp. 7–16.

Malik, K 2000, *Man, beast and zombies: what science can and can't tell us about human nature*, Weidenfeld and Nicolson, London.

Mazlish, B 1995, 'The man-machine and artificial intelligence', *Constructions of the Mind: Artificial Intelligence and the Humanities*, vol. 4, no. 2. Available from: <http://web.stanford.edu/group/SHR/4-2/text/mazlish.html>.

Meisner, M 1982, *Marxism, Maoism and utopianism: eight essays*, The University of Wisconsin Press, Madison.

Mori, M 1970, 'The uncanny valley', *Energy*, trans. KF MacDorman & T Minato, vol. 7, no. 2, pp. 33–35. Available from: http://www.androidscience.com/theuncannyvalley/proceedings2005/uncannyvalley.html accessed 28th December 2014.

Mori, M 1999, *The Buddha in the robot: a robot's engineers thoughts on science and religion*, Kosei Publishing Co., Tokyo.

Mori, M 2005, 'On uncanny valley', Proceedings Mukta Research Institute. Available from: http://www.androidscience.com/theuncannyvalley/proceedings2005/MoriMasahiro22August2005.html accessed 28th December 2014.

Mori, M 2012, 'The uncanny valley', *Robotics & Automation Magazine*, trans. KF MacDorman & N Kageki, vol. 19, no. 2, pp. 98–100.

Reichardt, J 1978, *Robots: fact, fiction and prediction*, Thames and Hudson Ltd., London.

Reilly, K 2011, *Automata and mimesis on the stage of theatre history*, Palgrave Macmillan, Basingstoke.

Ssorin-Chaikov, N 2006, 'On heterochony: birthday gifts to Stalin, 1949', *Journal of the Royal Anthropological Institute*, no. 12, pp. 355–375.

Standage, T 2002, *The mechanical Turk: the true story of the chess-playing machine that fooled the world*, Penguin, London.

Wood, G 2003, *Living dolls*, Faber and Faber, London.

Films Cited

2001: A Space Odyssey 1968, dir. Stanley Kubrick.

Blade Runner 1982, dir. Ridley Scott.

Conclusion
Loving the Attachment Wounded Robot

The relation to the Thou is direct. No system of ideas, no foreknowledge, and no fancy intervene between *I* and *Thou*.

Martin Buber, *I and Thou*, 1937, p. 11

In this book, I have tried to show that annihilation underscores the making of contemporary robotics and anthropological theorizing of the human and nonhuman relation. What is annihilated in the process? The social is annihilated because the social is the part of human existence where direct and loving interactions between one human and another are what can only make us human. The human is also annihilated: *the end of the human*. Annihilation anxiety is the anxiety expressed by rejecting categories of separation, if categories are not based on ontologically differences, then they are "reduced to nothing", "obliterated" and ontologically flattened, and therefore no meaningful essential difference can be reasonably argued, instead leaving a view of human and nonhuman agents as "assemblages" and "cyborgs", which is to say that they are presented as composites of multiple attachments to humans and nonhumans. This has led to a current trajectory in anthropological theorizing that is categorizing the field in terms of not what is separate or connected, but what is 'in-between space', or to coin Karen Barad's term, a realm of 'intra-objective becomings', where the authority of agency resides as an outcome of intra-objectivity (Barad 2003). The focus on the in-between space only leads to further entanglements about the authority of agency, because in the in-between space one is neither one thing or another, human and nonhuman at the same time (much like the robot). Analytically all categories are real and possible (dualisms, anti-dualistic positions, and paradoxes), though the content of such categories may change and shift over time. These shifts toward the in-between space are a reaction to the impossibility of a flat ontology, as theorists realize the impossibility of occupying more than one position, but the in-betweeners can do nothing to escape the nihilism of contemporary anthropological theorizing because at the foundation is a rejection of ontological difference.

For a robotic machine to be offered as a viable alternative to a human relationship shows considerable confusion about the nature of human attachments and how humans attach to one another. The robotic scientists who propose these robots to act as companions, therapists, lovers or friends are often motivated by a genuine desire to extend otherness into machines and to help humanity. Robotic machines and AI agents have become exemplars in showing us these new kinds of relational possibilities. In order for the AI and robotic machines to assume their ascendancy, a shift needs to occur in the modulation of what it means to be human. A human-robot attachment is only possible because of this mechanistic sociality that underscores contemporary sociality. The mechanical sociality is an outcome of an attachment crisis in how humans bond with others. Attachment wounds are an outcome of a lack of bond between one human and another and the attempts to use machines to help fill these gaps in social relations. Robot children, robot companions and robot therapists are the future!

If we return to one last tale from fiction to complete our journey, let us explore *AI: Artificial Intelligence* (2001). In the film a young robot child, David, is given to a family mourning the loss of a child who is in cybernetic hibernation while a cure can be found for his illness. In the interim, the family is given a robot child as a substitute. The child is designed with an imprinting code that, once recited, can never be erased. The mother, Monica, reads this code to David and activates his love imprint. A pleasant time occurs and their relationship grows: David takes pleasure in the story he is read of Pinocchio, such that one day he will have a wish to become a real boy. This time together is disturbed by the return of the family's son Martin. In time, Martin starts to dislike David, and Monica drives him to a field and disposes him there. She should send David back to the Mecha Corporation that created him to terminate him, but she cannot do this, so she leaves him in a field far from home. For the rest of his time, David searches for Monica—he searches to be reunited with the one that he loves. He takes the fiction of Pinocchio into himself and starts to use this fiction as his guide. He does not question if this fantasy is true or not, he just accepts that it is real. In turn, after a long and treacherous journey, David is discovered by an alien race. All humanity is long gone, and only the human-made remnants of humanity (things and machines) remain. They can use their science to bring Monica back to life for one day, and the last scenes of the film are of him spending one day with Monica.

The film is distressing and deeply sad, but this story reveals to us his separation anxiety, his profound feeling of disconnect. The machine, in the form of an embodied robot, will offer a connection to someone with an attachment wound. The machine in the form of a disembodied network also offers this connection. The machines work harder now to make humans feel connected to each other. But in accepting the mechanical robot, humanity must become more socially mechanical—less complex, more scripted, more stereotyped, and less spontaneous, adapting to the needs of the machine.

I will bring this book to a close by making final remarks on the themes of violence and love. Čapek's *R.U.R* (Rossum's Universal Robots) was the first fiction in which all humanity is destroyed. Not even in our tales of Frankenstein or the Golem, frightening tales about the limits of human agency, is all of humanity under threat. The play *R.U.R.* has been read as nonhuman (and later machines) becoming to powerful and taking over, but the source of the revolutionary agency in *R.U.R.* was drawn from human beings, not machines. Čapek wrote in the midst of revolutionary agency and labor struggles. In chapter one we explored how robots, imagined as human beings, were re-rendered as other artists as machines, and so as machine technologies develop, so too does a corresponding fear that they might rise up and become too powerful (this is still a theme today). This fear of the robots is really from another source. In the 1920s, violence, either from the battlefields or from revolution and political violence, were dominant themes. There were acts of violence to defend both capitalist and communist systems. By making the robots overthrow humanity, Čapek critiqued and rejected that that violence could bring about meaningful change—it can only lead to complete annihilation for humanity. Čapek's message is that any system that is built on violence can only end in violence: *it annihilates itself.* Violence destroys human bonds, it creates pain, anger, despair, sorrow, mourning and horror; no loving bond is possible through violence. In the last scene of the play, there are two robots that are in love, and there is hope that a new society can be built from this love. There is hope. Even in cases of the most serious human attachment difficulties, the attachment wound can be repaired by human love. The bond of love can be formed and the wound can be healed. Love begets more love.

BIBLIOGRAPHY

Barad, K 2003, 'Posthumanist performativity: toward an understanding of how matter comes to matter', *Signs: Journal of Women in Culture and Society*, vol. 28, no. 3, pp. 801–831.

Buber, M 1937, *I and thou*, trans. RG Smith, T. & T. Clark, Edinburgh.

Čapek, K 2004, *R.U.R. (Rossum's universal robots)*, Penguin Classics, New York.

Index